FUNDAMENTAL TOPICS IN

RELATIVISTIC FLUID MECHANICS AND MAGNETOHYDRODYNAMICS

FUNDAMENTAL TOPICS IN

RELATIVISTIC FLUID MECHANICS AND MAGNETOHYDRODYNAMICS

Edited by

Robert Wasserman

Charles P. Wells

Michigan State University

East Lansing, Michigan

Proceedings of a Symposium held at Michigan State University

October, 1962

1963

ACADEMIC PRESS · New York · London

ACADEMIC PRESS INC.
111 FIFTH AVENUE
NEW YORK 3, N. Y.

United Kingdom Edition
Published by
ACADEMIC PRESS INC. (LONDON) LTD.
BERKELEY SQUARE HOUSE, LONDON W. 1

Library of Congress Catalog Card Number: 63-22328

PRINTED IN THE UNITED STATES OF AMERICA

LIST OF CONTRIBUTORS

Peter Bergmann, *Department of Physics, Syracuse University, Syracuse, New York*

A. A. Blank, *Institute of Mathematical Sciences, New York University, New York, New York*

John Carstoiu, *Science for Industry, Incorporated, Brookline, Massachusetts*

Stephen Childress, *California Institute of Technology, Pasadena, California*

Tsung Lien Chou, *Michigan College of Mining and Technology, Houghton, Michigan*

N. Coburn, *Department of Mathematics, University of Michigan, Ann Arbor, Michigan*

M. Z. Krzywoblocki, *College of Engineering, Michigan State University, East Lansing, Michigan*

G.S.S. Ludford, *Department of Engineering Mechanics, Cornell University, Ithaca, New York*

Keith Leon McDonald, *Department of Physics, University of Utah, Salt Lake City, Utah*

E. W. Schwiderski, *U.S. Naval Weapons Laboratory Dahlgren, Virginia*

A. H. Taub, *Digital Computer Laboratory, University of Illinois, Urbana, Illinois*

T. Triffet, *Department of Metallurgy, Mechanics and Material Science, Michigan State University, East Lansing, Michigan*

John A. Wheeler, *Palmer Physical Laboratory, Princeton University, Princeton, New Jersey*

PREFACE

The mathematical formulation of physical problems has become more and more complex just as the problems themselves and their physical hypotheses have become more complex. In a sense one goal is to bring together what were once isolated parts of physics and merge them into a common but more general theory. Nowhere is this better illustrated than in the fields of fluid dynamics, electromagnetic theory, and relativity. The combining of classical fluid dynamics and relativity has been with us almost as long as the theory of relativity. However, the combination of fluid dynamics and electromagnetic theory has appeared as a branch of physics under the name of magnetohydrodynamics only in recent years.

Because of the rapid advances being made in both fields of relativistic fluid dynamics and magnetohydrodynamics it seemed desirable to bring together a group of those working in these fields for a symposium at which the more mathematical aspects would be presented. Accordingly, such a symposium was held at Michigan State University October 10-11, 1962. The published proceedings will provide a wide audience with the various aspects touched upon during the symposium. By its nature the proceedings will be of primary interest to those actively engaged in research in the fields of relativistic fluid dynamics and magnetohydrodynamics. However, it should also be of interest to those who have had training in any one of the individual fields and who wish to see how new results can be obtained by merging of the disciplines.

The organization of the volume essentially reflects that of the symposium. The first part is concerned with relativistic fluid dynamics and the second part devoted to magnetohydrodynamics. The first eight papers consist of the invited addresses and they are followed by five contributed papers. The number of contributed papers was purposely limited because of lack of time during the two-day meeting.

The symposium was restricted almost entirely to the mathematical aspects of the subject and this necessarily left out such topics as engineering applications (controlled fusion, space flight propulsion, etc.) and astrophysical and geophysical applications. Many symposia devoted to applications have taken place in recent years; e.g., the Midwest Conferences on Fluid Mechanics, the

University of Pennsylvania symposia, several international conferences on plasmas and ionization.

Our indebtedness to the authors of this volume is clear. In addition there are many others whose names do not appear here who deserve much credit. In particular there are the many members of the Mathematics Department, the Physics Department, the Division of Engineering Research, and several departments of the College of Engineering of Michigan State University who worked together to produce the symposium. To all these individuals and to all the others who actively participated in the meetings and helped to make the symposium a success we tender our sincere thanks.

ROBERT WASSERMAN
CHARLES P. WELLS

August, 1963

CONTENTS

SESSION OF CONTRIBUTED PAPERS

ABSENCE OF A GRAVITATIONAL ANALOG TO ELECTRICAL CHARGE

John Archibald Wheeler

Princeton University
Princeton, New Jersey

Charge and Topology

Curved empty space has been found to be a building material of unusual versatility. Out of it one can — in principle — construct a concentration of gravitational wave energy or electromagnetic wave energy or both (a "geon") which will hold itself together for an arbitrarily long time. One can also — again in principle — construct two electric charges of equal magnitude and opposite sign by trapping source-free electric lines of force in the topology of a doubly or multiply connected space.† Question: Can one construct a gravitational analog of electric charge by trapping source-free gravitational flux in a topological "handle" or "wormhole"?

Geometrodynamics

This question is analyzed here within the context of Einstein's general relativity, specialized to the case ("geometrodynamics") where no "real" matter is present. The only source considered for the gravitational field is electromagnetic

†For a summary of the developments of relativity theory mentioned in this paragraph, see Wheeler [1].

stress, energy, and energy flow. The source-free Maxwell equations plus Einstein's field equations define the dynamical system under consideration, except for boundary conditions. This system will be analyzed here only at the classical level. At this level the geometrodynamical objects which present themselves — geons and charges — however interesting, and however reminiscent in some respects of the real physical world, have not the slightest *direct* connection with elementary particles. Therefore the present investigation seeks only to spell out in a little more detail some of the implications of classical standard general relativity theory. It is not the expectation to gain any insight into the much deeper problems of strangeness and baryon number.

The conclusion from the investigation is simple and negative: there is no classical gravitational analog of the classical type of electrical charge that shows up in a multiply connected space.

Why ever suppose in the first place that there might have been a gravitational analog for electric charge? Because there are so many analogies between the initial value problems of electrodynamics and geometrodynamics [2].

Initial Value Problem for Electrodynamics

One cannot specify the initial value problem for the source-free electromagnetic field in the two-surface formulation without stating what is the flux of the electric field through each handle of the topology. Specifically, give the magnetic field $\mathbf{B}'(x,y,z)$ on one space-like hypersurface σ' and the magnetic field $\mathbf{B}''(x,y,z)$ on a second space-like hypersurface σ''. Specialize for simplicity to the case where the two hypersurfaces are so close together that one can more appropriately speak of a single hypersurface σ on which one knows everywhere $\mathbf{B}$ and $\partial\mathbf{B}/\partial t$. It is assumed that

$$\operatorname{div} \mathbf{B} = 0 \tag{1}$$

and

$$(\partial/\partial t) \operatorname{div} \mathbf{B} = \operatorname{div} (\partial\mathbf{B}/\partial t) = 0. \tag{2}$$

This condition implies that *locally* $\mathbf{B}$ can be represented as the curl of a magnetic potential $\mathbf{A}$. Moreover, it is assumed that there is no flux of magnetic lines of force through any handle or wormhole of the topology. In other words, it is assumed that there is no classical magnetic charge. This circumstance implies [3] that not only locally but also *globally* throughout the hypersurface one can define a single-valued vector field $\mathbf{A}$ — and associated time rate of change $\partial \mathbf{A}/\partial t$ — such that

$$\mathbf{B} = \operatorname{curl} \mathbf{A} \tag{3}$$

$$\partial \mathbf{B}/\partial t = (\partial/\partial t) \operatorname{curl} \mathbf{A} = \operatorname{curl} (\partial \mathbf{A}/\partial t) . \tag{4}$$

The "initial value problem" requires one to find on the initial hypersurface σ an electric field $\mathbf{E}(x, y, z)$ such that

$$\operatorname{div} \mathbf{E} = 0 \tag{5}$$

$$\operatorname{curl} \mathbf{E} = -\partial \mathbf{B}/\partial t . \tag{6}$$

The initial value problem has to be solved before one can take up properly the dynamic problem, to project $\mathbf{B}$ and $\mathbf{E}$ off the hypersurface and predict the entire past and future of the electromagnetic field. The initial value problem has a unique solution

$$\mathbf{E} = \mathbf{E}_{\text{unique}}(x, y, z) \tag{7}$$

in a differentiable closed manifold, no matter how irregularly it may be curved, provided that has the topology of the 3-sphere, S^3. In a multiply connected closed space, endowed with n handles ($n = R_2$ = second Betti number of the manifold) the electric field is no longer uniquely determined solely by the magnetic field and its time rate of change. It is only fully fixed when one specifies in addition n charge-like parameters q_k. Each parameter measures the flux of the electric field through one of the handles. The parameter q_k multiplies a harmonic vector field $\mathbf{H}_k$ with the following properties:

$$\operatorname{div} \mathbf{H}_k = 0 \tag{8}$$

$$\operatorname{curl} \mathbf{H}_k = 0 \tag{9}$$

$$\int_{\{c_j\}} \mathbf{H}_k = 4\pi\,\delta_{jk}\,. \tag{10}$$

Here the surface integral is extended over the mouth of the jth handle — or, in sharper language, over one of the 2-cycles which belongs to the jth independent homology class. By way of interpretation, consider the case where the mouth of a handle or wormhole is very small and the radius of curvature of the surrounding space is very large. Then the harmonic vector field associated with this wormhole looks like the Coulomb field of a point charge except in the immediate vicinity of the mouth. The harmonic vector fields associated with other handles have no flux through this particular handle. In terms of these $n = R_2$ independent harmonic vector fields the solution of the initial value problem is

$$\mathbf{E}(x,y,z) = \mathbf{E}_{\text{unique}} + \sum_k q_k \mathbf{H}_k\,. \tag{11}$$

Thus, to be able to predict the past and future of the field one must give, in addition to $\mathbf{B}$ and $\partial\mathbf{B}/\partial t$, the classical charge-like magnitudes q_k.

By analogy, one might suppose that also in geometrodynamics it is not enough to specify the intrinsic 3-geometry and its time rate of change; that one must give in addition the values of certain charge-like parameters to determine the field momentum — and thus to fix the entire past and future evolution of the geometry. However, this supposition — it is concluded here — is false (Table I).

Variational Principle

The analysis in both cases is carried out more conveniently in terms of potentials than in terms of the gage-independent physical quantities themselves; and more conveniently in terms of the appropriate variational principle than in terms of the equivalent differential equations. In electrodynamics, write

$$\mathbf{E} = -\partial\mathbf{A}/\partial t - \operatorname{grad}\varphi\,. \tag{12}$$

TABLE I
ANALOGY BETWEEN ELECTRODYNAMICS AND GEOMETRODYNAMICS PURSUED TO THE POINT WHERE IT BREAKS DOWN

The two-surface formulation of the initial value problem in a closed and multiply connected space of second Betti number R_2 requires one in the case of electrodynamics to specify $n = R_2$ classical charge-like parameters, but there is no place for any such parameters in the case of geometrodynamics. The quantities $T^*_{\perp\perp}$ and $T^*_{i\perp}$ represent (G/c^4) multiplied, respectively, by the energy density and by the ith component of the flow of energy. The extrinsic curvature tensor, transformed to diagonal form, gives three numbers. A given one of these numbers represents the fractional shrinkage of a vector located in the hypersurface when its two ends are transported a unit distance perpendicular to the hypersurface. The subscript $|j$ represents covariant differentiation with respect to the jth coordinate *in* the hypersurface, this covariant differentiation being defined with respect to the *three*-dimensional metric (in contrast to $;j$ which would denote covariant differentiation with respect to the 4-metric).

	Electrodynamics	Geometrodynamics
The physically significant data that one thinks of specifying on two space-like hypersurfaces	The magnetic field on each hypersurface	The 3-geometry intrinsic to each hypersurface
Are these data gage independent?	Yes	Yes

TABLE I (Continued)

	Electrodynamics	Geometrodynamics
A gage-dependent way of specifying these data	Give $\mathbf{A}'(x,y,z)$ and $\mathbf{A}''(x,y,z)$	Give ${}^{(3)}g'_{ik}(x,y,z)$ and ${}^{(3)}g''_{ik}(x,y,z)$
Nature of gage transformation	$\mathbf{A} \rightarrow \mathbf{A} + \operatorname{grad} \lambda$ (one gage function λ)	${}^{(3)}g_{ik} \rightarrow {}^{(3)}g_{mn} \dfrac{\partial x^m}{\partial \bar{x}^i} \dfrac{\partial x^n}{\partial \bar{x}^k}$ (change in 3 space coordinates)
Sharpening of initial value data to case of two hypersurfaces at an infinitesimal separation	Give $\mathbf{A}(x,y,z)$ and $\partial \mathbf{A}/\partial t$ on the limiting single hypersurface	Give ${}^{(3)}g_{ik}(x,y,z)$ and $\partial\, {}^{(3)}g_{ik}/\partial t$ on the limiting single hypersurface
Having the field "coordinate" and its time rate of change, what does one require in order to have the field *momentum*, so as to be able to predict the entire past and future?	The electric field $\mathbf{E}(x,y,z)$ on the initial space-like hypersurface	The extrinsic curvature $K_{ik}(x,y,z)$ telling how the hypersurface is curved relative to the surrounding and yet-to-be-constructed 4-geometry

TABLE I (Concluded)

	Electrodynamics	Geometrodynamics
Initial value equations to be used to determine the foregoing quantity	$\operatorname{div} \mathbf{E} = 0$ $\operatorname{curl} \mathbf{E} = -\partial \mathbf{B}/\partial t$	$(K^j_i - \delta^j_i K)_{\mid j} = 16\pi T^*_{\perp i}$ ${}^{(3)}R + (\operatorname{Tr} \mathbf{K})^2 - \operatorname{Tr} \mathbf{K}^2 = 16\pi T^*_{\perp\perp}$
Are the given data enough to determine the solution of these equations in a closed but multiply connected 3-manifold?	No; must specify in addition $n = R_2$ classical charge-like parameters q_k if one is to fix $\mathbf{E}$ uniquely	Yes; no place for specification of parameters which might represent some gravitational analog of electric charge (the result of present paper)

Then one of the initial value equations (5, 6) is automatically satisfied. The other becomes a single equation for a single unknown:

$$\operatorname{div}\operatorname{grad}\varphi = -\operatorname{div}(\partial\mathbf{A}/\partial t). \tag{13}$$

This equation flows straight out of the variational principle

$$(1/8\pi)\int \mathbf{E}^2\, d(\text{volume}) = \text{extremum} \tag{14}$$

when φ is regarded as the quantity to be varied to achieve the extremum ($\partial\mathbf{A}/\partial t$ being already specified).

This variational principle requires a close discussion in the case of a closed and orientable [4] but multiply connected space. One is demanding that φ be so chosen as to annul the first order variation

$$(1/4\pi)\int \mathbf{E}\cdot\delta\mathbf{E}\, d(\text{volume}) = -(1/4\pi)\int \mathbf{E}\cdot\operatorname{grad}\delta\varphi\, d(\text{volume}). \tag{15}$$

An integration by parts gives

$$(1/4\pi)\int \delta\varphi\operatorname{div}\mathbf{E}\, d(\text{volume}) - (1/4\pi)\int \delta\varphi(\mathbf{E}\cdot d\mathbf{S}) = 0\,. \tag{16}$$

The potential φ can be varied freely anywhere within the volume. Therefore the coefficient of $\delta\varphi$, the quantity $\operatorname{div}\mathbf{E}$, must vanish everywhere. So far there is nothing with which one is not familiar from the corresponding problem in Euclidean space. However, now comes the quantity to be evaluated at the limits.

Conceive of the volume integration as proceeding outward from some chosen point and engulfing more and more volume. The front where new contributions are being added in may be compared with the 2-dimensional surface of a balloon that is being inflated in 3-dimensional space. If the space has the topology of the 3-sphere $\mathbf{S}^3$, the balloon ultimately shrinks to nothingness at some other point in the closed space.† In the

†By way of analogy, consider an example with one less dimension: a 2-sphere or globe. On this surface a rubber band that starts expanding about the north pole reaches a maximum extension at the equator and shrinks to nothingness at the south pole.

case of a multiply connected space, the balloon surface bulges down more and more into each wormhole mouth. Somewhere in the throat of a wormhole the two expanding rubber surfaces at last meet and the integration as regards that part of space is at last complete. There are as many interfaces of this kind as there are handles in the space. At each such interface there is a *pair* of contributions to the surface integral in (16).

The orientability of the manifold means that the right-hand rule is agreed upon as between a man behind one of the rubber surfaces and another man behind the rubber surface that confronts it. However, the thumb of each — the outward drawn normal — points toward the other. Consequently the values of $(\mathbf{E} \cdot d\mathbf{S})$ for the two sheets for a given small portion of the interface are equal and opposite. Moreover $\delta\varphi$ is the same on both sheets. Therefore all surface contributions vanish.

Parameters Needed to Define Variational Problem

To say that $\delta\varphi$ is the same on one sheet as on the other is not the same as saying that φ itself is the same on the two sides of the interface. The physically significant quantity is the electric field, not the potential. The electric field $\mathbf{E}$ is single valued. The contribution $-\partial\mathbf{A}/\partial t$ to the electric field, $-\text{grad}\ \varphi$, is also single valued. It follows that φ itself can change across the interface by a constant — but only by a constant. *The value of this constant has to be specified before the variational problem is well defined.*

There are as many such constants to be fixed in advance as there are handles in the multiply connected space. None of these constants is itself a charge-like parameter q_k. Instead, it is in a certain sense the conjugate of a charge, namely the potential change $\Delta\varphi_k$ associated with the traversal of the corresponding wormhole. There is a linear relation between the potential changes and the charges, of the form

$$\begin{aligned} \Delta\varphi_1 &= C_{11}q_1 + \dots + C_{1n}q_n \\ &\dots\dots\dots\dots\dots \\ \Delta\varphi_n &= C_{n1}q_1 + \dots + C_{nn}q_n\,. \end{aligned} \tag{17}$$

Whether one specifies the n charges q_k or the n potential changes $\Delta\varphi_k$ or n independent combinations of these quantities, one cannot escape having to fix n quantities over and above the values of $\mathbf{A}$ and $\partial \mathbf{A}/\partial t$ in order to provide a well-defined initial value problem.

If there were two solutions φ_A and φ_B of the initial value problem associated with the *same* $\partial \mathbf{A}/\partial t$ and with the *same* $\partial A/\partial t$ and with the *same* interface jumps $\Delta\varphi_1, \ldots, \Delta\varphi_n$, then their difference

$$u = \varphi_B - \varphi_A \tag{18}$$

would have zero jump at each interface and thus be a single valued solution of the equation

$$\text{div grad } u = 0 \ . \tag{19}$$

But the only bounded solution of this equation in a closed space of any connectivity is

$$u = \text{constant} \ . \tag{20}$$

The argument is familiar. The boundedness of u implies either (1) that u is everywhere constant or (2) there is some point or region where u has a maximum value $u_{\max}$ greater than the value assumed by u elsewhere. But (2) is impossible. It implies that there exists at least one point P at which u takes on a value greater than the average of the values of u over the surface of a small sphere centered on P. Contrary to this, Eq. (19) says that the average of u over the surface of the small sphere centered on P is *equal* to the value of u at P itself. Therefore u must be constant. Consequently the solution E of the initial value problem is uniquely specified by $\partial \mathbf{A}/\partial t$ and the n quantities $\Delta\varphi_k$.

Initial Value Problem for Geometrodynamics

It only remains to go through the same reasoning for the initial value problem of geometrodynamics and see where it goes wrong. For simplicity, the sources are treated as given; no attempt is made to treat the initial value problems of geometry and electromagnetism simultaneously. Therefore the energy density $T^*_{\perp\perp}(x,y,z)$ and the energy flow $T^*_{i\perp}(x,y,z)$ are regarded as specified functions of position. Given also is the 3-geometry ${}^{(3)}g_{ij}(x,y,z)$ and its rate of change with respect to a "time" parameter, $\partial\,{}^{(3)}g_{ik}(x,y,z)/\partial t$, on a hypersurface arbitrarily called t = constant. The problem is to find the physical quantity which will complete the initial value data: to evaluate the extrinsic curvature $K_{ij}(x,y,z)$ ("second fundamental form") of the hypersurface with respect to the surrounding — and yet to be found — four-space. As in electromagnetism, so here it is easier to solve for potentials from which K_{ij} can be found that it is to solve directly for this extrinsic curvature itself.

Following Arnowitt, Deser and Misner, write the geometry of 4-space near the hypersurface in question in the form

$$d\sigma^2 = -d\tau^2 = g_{\alpha\beta}\, dx^\alpha\, dx^\beta$$

$$= {}^{(3)}g_{ik}\, dx^i\, dx^k + 2N_i\, dx^i\, dx^0 + (N_i N^i - N_0^2)(dx^0)^2 \quad (21)$$

Here N_0 is the lapse function [5]: the *proper* time separation between two hypersurfaces — measured along a normal — per unit of difference in the time *coordinate* t. The vectorial *shift* function [6] N^i represents the ith space coordinate at the base of the normal diminished by the coordinate at the summit of the normal, this difference again being referred to a unit difference between the time *coordinates* of the two hypersurfaces. In terms of the four potentials N_0 and $N_i = {}^{(3)}g_{ik}N^k$ the extrinsic curvature of the hypersurface as it is to be imbedded in a yet-to-be-constructed 4-geometry is

$$K_{ij} = (1/2N_0)(N_{i|j} + N_{j|i} - \partial\,{}^{(3)}g_{ij}/\partial t)\,. \quad (22)$$

This equation is the analog of the expression (12) for the electric field in terms of the gradient of the scalar potential φ (the analog of the four N_α) and the time rate of change of the vector potential (the analog of the six ${}^{(3)}g_{ij}$).

The two-surface formulation of the initial value problem [2] follows the lead of the two-surface formulation of electrodynamics. One specifies the six potentials ${}^{(3)}g_{ij}$ on two hypersurfaces or more conveniently these potentials and their rate of change with respect to a time parameter on one hypersurface. One then seeks to find the four potentials — the lapse function N_0 and the vectorial shift function N^i — from the four initial value equations. One of these equations (the last in Table I) is immediately solved to give the lapse between successive hypersurfaces:

$$N_0 = +\left[\gamma_2\left(2T^{**}_{\perp\perp} - {}^{(3)}R\right)\right]^{1/2} \tag{23}$$

Here γ_2 is the *shift anomaly*

$$\gamma_2 = (\mathrm{Tr}\,\gamma)^2 - \mathrm{Tr}\,\gamma^2 , \tag{24}$$

where

$$\gamma_{ij} = N_0 K_{ij} = \frac{1}{2}\left(N_{i|j} + N_{j|i} - \partial\,{}^{(3)}g_{ij}/\partial t\right) . \tag{25}$$

Then one is left with three "condensed" initial value equations for the remaining three undetermined potentials, the three components N^i of the shift of one hypersurface relative to a nearby one:

$$\left\{\frac{\left(2T^{**}_{\perp\perp} - {}^{(3)}R\right)^{1/2}\left(\gamma_{ij} - {}^{(3)}g_{ij}\,{}^{(3)}g^{mn}\gamma_{mn}\right)}{2\left({}^{(3)}g^{ab}\,{}^{(3)}g^{cd} - {}^{(3)}g^{ac}\,{}^{(3)}g^{bd}\right)\gamma_{ab}\gamma_{cd}}\right\}^{|j} = -T^{**}_{\perp i}$$

$$= +\,(8\pi\,G/c^4)\begin{pmatrix}i\text{th covariant component of}\\ \text{density of flow of energy}\end{pmatrix} . \tag{26}$$

Let the initial value problem now be reviewed. One specifies:

(1) the topology of a 3-space. This topology determines how many coordinate patches are required to cover the space without singularity. In each coordinate patch — which is to overlap slightly on its neighbors — one lays down

(2) a nonsingular coordinate system. One next specifies an intrinsic geometry ${}^{(3)}G$ on the manifold, and translates this in each coordinate patch into

(3) a specification of the six ${}^{(3)}g_{ij}(x, y, z)$. Similarly one gives as a function of the three coordinates:

(4) the six ${}^{(3)}\dot{g}_{ij} = \partial\,{}^{(3)}g_{ij}/\partial t$, where t is a parameter that is not otherwise defined. In other words, t is only a bookkeeping device to keep one hypersurface distinct from another. Finally one gives over the 3-manifold:

(5) the energy density or more conveniently $(8\pi G/c^4)$ times this energy density, a quantity with the dimensions of the inverse square of a length and designated here as $T^{**}_{\perp\perp} = -T^{**\perp}_{\perp} = T^{**\perp\perp}$; and

(6) the multiple $(8\pi G/c^4)$ of the density of flow of energy, denoted here by $T^{**\perp}_{i} = -T^{**}_{i\perp}$.

Then the condensed initial value equations (26) become well defined. One desires

(1) to solve† for the three components N_i of the shift; then
(2) to calculate the lapse N_0 from the simple Eq. (23); then
(3) to evaluate the extrinsic curvature K_{ij} from the equation of definition (22).

Then one has enough well defined initial value data so that one can

(4) predict the evolution of the geometry in past and future.‡

Requirements Needed to Guarantee Uniqueness

This program will not lead to a uniquely defined 4-geometry unless the three differential equations for the three components of the shift have a unique solution. The solution will not be unique if there is no boundary condition to distinguish one solution of the condensed initial value equations from another. Such a boundary condition is not contained in Einstein's field equations. It has to be added to those equations. Only then can one have a properly complete formulation of general relativity as a physical theory.

---

†The solution for a special case of the general problem is given in MP [6].

‡It is assumed here that the initial value problem for the Maxwell field is carried along and treated in parallel with that for the geometry, so that the two sets of second order equations can be integrated as a coupled system. It is compatible with this assumption to treat among others the special case where the electromagnetic field vanishes on the hypersurface. Then the field is zero at all times and $T^{**}_{i\perp}$ and $T^{**}_{\perp\perp}$ vanish.

In a recent report [7] it is argued that the appropriate boundary condition is this, that the 3-dimensional hypersurface should be closed. Reasons are given for thinking that this closure condition, plus the initial data [1-6], determine the N_i, and hence N_0, the K_{ij}, and the entire 4-geometry. Such a determination of the 4-geometry fixes the course of all geodesics. Then the inertial properties of every infinitesimal test particle are directly connected back to the distribution of energy and energy flow ($T^{**}_{\perp\perp}$ and $T^{**}_{i\perp}$) and of effective gravitational wave energy (as specified by ${}^{(3)}g_{ij}$ and $\partial{}^{(3)}g_{ij}/\partial t$) throughout the 3-manifold. The principle of Mach, connecting inertia with the distribution of mass throughout space, acquires in this way not only a definite place in the "plan" of general relativity, but also a well defined mathematical formulation.

In connection with the foregoing formulation of Mach's principle the question was raised whether it is enough to demand closure and to specify the data [1-6] in order to find the remaining four potentials N_a. Will one not have also to specify in geometrodynamics as he does in electrodynamics some charge-like parameters? Short of such a specification can one expect that the listed data will be sufficient to define a unique solution for the dynamical problem?

This question of charge, it now appears, is most easily treated in geometrodynamics as in electrodynamics by going to a variational principle that is equivalent to differential equations *plus* boundary condition. Such a variational principle was given in the cited report but the idea was not then evident that it could be applied to the desired end. This variational principle is a simple generalization — to the case where energy and energy flow are present — of a variational principle developed by Ralph Baierlein and reported in reference [2]. This reference in turn is based on the two-surface formulation of relativity as given by David Sharp [8]. The generalized variational principle reads

$$I = \int \left\{ \left[\gamma_2 \left(2T^{**}_{\perp\perp} - {}^{(3)}R \right) \right]^{1/2} + T^{**k}_{\perp} N_k \right\} \left({}^{(3)}g \right)^{1/2} d^3x = \text{extremum.} \quad (27)$$

The variational principle has the virtue that it tells directly what has to be specified to provide a definite solution for the three N_i. Let the N_i be varied. Let any term which contains the derivative of a δN_i be integrated by parts so that this quantity appears undifferentiated. One ends up as in electrodynamics with a volume term and surface terms. Demand that the volume term vanish for arbitrary δN_i. In this way one comes to the condensed initial value equations (26) — so far nothing not already reported [7]! Now for the surface terms! It is worthwhile to outline the derivation of these terms.

From the definition

$$\gamma_{ij} = \frac{1}{2}(N_{i|j} + N_{j|i} - \dot{g}_{ij}) \tag{28}$$

one finds the variation

$$\delta\gamma_{ij} = \frac{1}{2}(\delta N_{i|j} + \delta N_{j|i}) . \tag{29}$$

From this result and from the definition of the shift anomaly (24) one finds the variation

$$\delta\gamma_2 = 2\left({}^{(3)}g^{ij}\,\mathrm{Tr}\,\boldsymbol{\gamma} - \gamma^{ij}\right)\delta N_{i|j} . \tag{30}$$

Taking the variation of the I of Eq. (27) and substituting in this expression for $\delta\gamma_2$, one obtains

$$\begin{aligned}\delta I &= \int\left\{\left[\left(2T^{**}_{\perp\perp} - {}^{(3)}R\right)\Big/\gamma_2\right]^{1/2}\left(g^{ij}\,\mathrm{Tr}\,\boldsymbol{\gamma} - \gamma^{ij}\right)\delta N_{i|j}\right.\\ &\quad\left. + T^{**k}_{\perp}\,\delta N_k\right\}\left({}^{(3)}g\right)^{1/2} d^3x\\ &= \int\left\{\left(g^{ij}\,\mathrm{Tr}\,\mathbf{K} - K^{ij}\right)\delta N_{i|j}\right.\\ &\quad\left. + T^{**k}_{\perp}\,\delta N_k\right\}\left({}^{(3)}g\right)^{1/2} d^3x .\end{aligned} \tag{31}$$

Integrate by parts and find the surface term

$$\int(g^{ij}\,\mathrm{Tr}\,\mathbf{K} - K^{ij})\,\delta N_i\, d\mathbf{S}_j \tag{32}$$

in analogy to the surface term

$$\int E^j\,\delta\varphi\, d\mathbf{S}_j \tag{33}$$

in electromagnetic theory.

No Extrinsic-Geometry Analog to Electric Charge

Both in the gravitational problem and in the Maxwell problem each wormhole throat — or in better mathematical language, each independent homology class — presents one with a pair of surfaces. The physically meaningful quantity — in one case the electric field, in the other case a construct out of the extrinsic curvature tensor — is continuous across the double-layer interface. To secure the vanishing of the surface term one demands in the one case that $\delta\varphi$, in the other case that δN_i, shall have the same value on the two sides of the interface. In the electromagnetic case this requirement did *not* imply that φ itself should be continuous across the interface. On the contrary, it was necessary to specify the jump $\Delta\varphi$ in φ across the interface to make the variational problem well defined in a compact multiply connected space. Out of this circumstance followed the existence of as many charge-like parameters as there are independent classes of homologous closed 2-surfaces. But now comes the difference. In the gravitational case not only δN_i, but also N_i itself, must have the same value on the two sides of the interface. There is no room for the specification of further independent parameters. This follows from the circumstance that the geometry (21) near the hypersurface has to be single valued. It would not make sense to have a jump in N_i. Such a jump would imply the existence of two different values for the distance between two *nearby* points (x,y,z,t) and $(x+dx,\ y+dy,\ z+dz,\ t+dt)$. A double-valuedness in this distance would be contrary to the basic postulate of general relativity that *locally* spacetime has the standard Lorentz character. Hence the conclusion which is the point of this paper: *there exists in the geometrodynamics of a closed multiply connected space no* extrinsic *geometric analog to the classical wormhole type of electric charge.*

The Conjugate Formulation of the Initial Value Problem

What this conclusion leaves unsettled is as important to state as what it settles. It says nothing about the existence or nonexistence of an *intrinsic* geometric analog to trapped electric lines of force. By way of motivating this distinction between extrinsic geometry ("geometrodynamical field momentum") and intrinsic geometry ("geometrodynamical field coordinate"), turn to the initial value problem of electrodynamics. Consider this problem in Sharp's two-surface formulation, but in the representation *conjugate* to that employed in this paper. Give on each of the two nearby space-like hypersurfaces, not the field "coordinate" **B**, but the field "momentum" **E**; or more conveniently, give on one hypersurface the electric field and its time rate of change. Then it is not necessary or even possible to specify independently — as part of the initial value data — the classical charge-like quantities q_k. The electric flux through each handle of a multiply connected space is *already* determined by the electric field itself. Thus, if one counted only the number of quantities to be specified, and worked only in the momentum representation, he might mistakenly conclude that there is no place for the concept of charge in source-free electromagnetism.

A closer look at the representation in which one specifies **E** and $\partial\mathbf{E}/\partial t$ on a space-like hypersurface is appropriate here for the questions it will raise about the corresponding representation of the initial value problem of geometrodynamics. Maxwell's equations are well known to be invariant with respect to the substitution

$$\mathbf{E} \to \mathbf{B}$$

$$\mathbf{B} \to -\mathbf{E}. \tag{34}$$

Why then should it make any difference whether one specifies on the hypersurface **B** and $\partial\mathbf{B}/\partial t$ or **E** and $\partial\mathbf{E}/\partial t$? Must not the initial value data fulfill in both representations the same type of initial value condition:

$$\operatorname{div}\mathbf{B} = 0, \quad \operatorname{div}(\partial\mathbf{B}/\partial t) = 0 \tag{35}$$

or

$$\operatorname{div}\mathbf{E} = 0, \quad \operatorname{div}(\partial\mathbf{E}/\partial t) = 0? \tag{36}$$

Yes; but there is more to the story. In one way of telling it, the field is assumed to admit of a global representation in terms of the usual kind of 4-potential,

$$\begin{aligned}\mathbf{B} &= \operatorname{curl}\mathbf{A}\\ \mathbf{E} &= -\partial\mathbf{A}/\partial t - \operatorname{grad}\varphi,\end{aligned} \tag{37}$$

but is not in general assumed to admit of a global representation of the converse character,

$$\mathbf{E} = \mathrm{curl}\,\mathbf{C}\,,$$
$$\mathbf{B} = \partial\mathbf{C}/\partial t + \mathrm{grad}\,\psi\,. \qquad (38)$$

Why? No magnetic poles are known in the real world of quantum physics. Therefore it is assumed that already in the classical theory any net flux of magnetic field through any handle of the topology is forbidden. This is the condition for the global existence of the magnetic potential **A**. In contrast, flux of *electric* lines of force through a wormhole is assumed to be allowed. However, if **E** could be written as the curl of a globally defined electric vector potential **C**, the surface integral of **E** over a sphere surrounding the mouth of a wormhole would vanish automatically. There could be no electric flux through the wormhole, contrary to assumption. Therefore the second way of describing the electric field (Eq. 38) is regarded as nonphysical. In contrast, the first representation of the electromagnetic field (Eq. 37) is accepted. The quantities **A** and φ are considered as having physical meaning, apart from a gage transformation (Eq. 38). In other words, the representation of the electromagnetic field in terms of the usual 4-potential is founded in the last analysis upon *observation.*†

One is open to believing on this account that an analysis of the existence or nonexistence of a geometrodynamical analog to trapped electric lines of force must also go back to experience for a proper foundation. We demand that the geometrodynamical potentials, the lapse function N_0 and the shift functions N_i, be *single valued* functions of position. We make this demand because we count the geometry of space-time — unlike the electromagnetic potentials — as directly observable (tracks of world lines and light rays, etc.). Therefore in a certain sense the existence of a geometrodynamical type of charge was not disproven mathematically, but ruled out on what seems to be a reasonable physical argument.

†See for example L. Witten's construction of examples in which arbitrary mixtures of electric and magnetic fields are trapped in the topology, because — deliberately, for the sake of experimentation with theory — any representation of the form (37) for the fields in terms of a 4-potential was given up.

Any Intrinsic-Geometry Analog to Electric Charge?

Two courses seem open to anyone determined to investigate to the bitter end the possibilities for geometrodynamical charge within the framework of Einstein's standard 1915 relativity theory. (1) Assume that the geometrical potentials N_α are not single-valued functions of position at the scale of distances $\sim(hG/c^3) = 1.6 \times 10^{-33}$ cm characteristic of quantum phenomena, however much they give the appearance of single valuedness at the scale of nuclear and atomic physics and larger dimensions. Not the slightest reason is evident to suggest such a radical break with well established thinking. Or (2), look at the dynamically conjugate representation of the 2-surface initial value problem. Here — in the limit of two hypersurfaces indefinitely close together — one specifies on one hypersurface the geometrodynamical *field momentum*

$$\pi^{ij} = g^{1/2}\left({}^{(3)}g_{ij}K - K_{ij}\right), \tag{39}$$

$$K_{ij} = -\left(\frac{1}{2}N_0\right)(\dot{g}_{ij} - N_{i|j} - N_{j|i}), \tag{40}$$

with its time rate of change, $\dot{\pi}^{ij}$. With *this* initial data the 3-geometry (the "field coordinate") is not known but has to be found. How many regular and independent solutions of *this* initial value problem are there in a manifold with a specified topology? Is the intrinsic 3-geometry unique? Or is the solution arbitrary to the extent of a number of adjustable parameters — this number being fixed by the topology? If so, these parameters would be called the geometrodynamical analog of electric charge. In brief, the present analysis shows that — *as regards the question of a new kind of charge* — one has to give up the following proposed analogy between electromagnetism and relativity:

Specified	~Thereby determine (up to R_2 adjustable charge-like parameters)
B, $\dot{\mathbf{B}}$	**E**
${}^{(3)}g_{ij}$, ${}^{(3)}\dot{g}_{ij}$	π^{ij} (wrong; is *unique*)

Only an investigation of the conjugate initial value problem will show if geometrodynamical charge is also excluded in the following trial correspondence:

Specified	Thereby determine (up to R_2 adjustable charge-like parameters)
$\mathbf{B}$, $\dot{\mathbf{B}}$	$\mathbf{E}$
π^{ij}, $\dot{\pi}^{ij}$	${}^{(3)}g_{ij}$ (unique?)

REFERENCES

[1] J. A. Wheeler, "Geometrodynamics." Academic Press, New York, 1962 (referred to hereafter as GMD).
[2] R. F. Baierlein, D. H. Sharp, and J. A. Wheeler, *Phys. Rev.* 126, 1864 (1962).
[3] GMD, p. 273 ff.
[4] GMD, p. 272.
[5] R. Arnowitt, S. Deser, and C. W. Misner, *Phys. Rev.* 122, 997 (1961) and earlier papers cited by them.
[6] The significance of the lapse and shift functions is discussed in J. A. Wheeler, "Mach's principle as boundary condition for Einstein's field equations," report given at Conference on relativistic theories of gravitation, Warsaw, July 25-31, 1962, as adapted for publication in "Lectures in Theoretical Physics," Vol. V (W. E. Brittin and B. W. Downs, eds.). Interscience, New York (1962 lectures published in 1963); also to appear in "Gravitation and Relativity" (H. Y. Chiu and W. F. Hoffman, eds.). Benjamin, New York, 1963 (cited hereafter as MP).
[7] MP, reference 6.
[8] David Sharp, A.B. Senior Thesis, Princeton University, May 1960, unpublished.

HYDRODYNAMICS AND GENERAL RELATIVITY

A. H. Taub

University of Illinois
Urbana, Illinois

1. Introduction

It may be shown that the general relativistic theory of the behavior of a physical system described by a tensor field with components φ^A and whose motion is governed by a variational principle

$$\delta \int \mathcal{L}(\varphi^A, \varphi^A_{;\sigma}) \sqrt{-g}\, d^4x = 0$$

is contained in the closed system of equations

$$R^{\mu\nu} - \frac{1}{2} g^{\mu\nu} R = -kc^2 T^{\mu\nu} \tag{1.1}$$

$$F_A \equiv \frac{\partial \mathcal{L}}{\partial \varphi^A} - (p^\sigma_A)_{;\sigma} = 0 \tag{1.2}$$

where the semicolon denotes covariant differentiation with respect to the metric $g_{\sigma\tau}$ and

$$p^\sigma_A = \frac{\partial \mathcal{L}}{\partial \varphi^A_{;\sigma}} \tag{1.3}$$

and $T^{\mu\nu}$ is expressed as a function of the φ^A and the $\varphi^A_{;\sigma}$ (or p_A^{σ}).

The theory of relativistic hydrodynamics is concerned with the case where $T^{\mu\nu}$ is of a particular form, namely,

$$T^{\mu\nu} = \sigma U^{\mu} U^{\nu} - \frac{p}{c^2} g^{\mu\nu} . \tag{1.4}$$

It is degenerate in the sense that the equations $F_A = 0$ are equivalent to

$$T^{\mu\nu}_{;\nu} = 0 \tag{1.5}$$

which are a consequence of (1.1). Thus there are no equations for the determination of the dependent variables describing the state of the fluid other than the Einstein field equations themselves.

2. Algebraic Characterization of $T^{\mu\nu}$

The four-vector field U^{μ} occurring in Eq. (1.4) will be called the four-velocity vector of the fluid. We require it to be a time-like proper vector of the tensor T^{μ}_{ν} and normalize it to satisfy

$$g_{\mu\nu} U^{\mu} U^{\nu} = 1 \tag{2.1}$$

$$T^{\mu}_{\nu} U^{\nu} = \rho(1 + \epsilon/c^2) U^{\mu} . \tag{2.2}$$

Thus we write

$$\sigma = \rho(1 + \epsilon/c^2 + p/\rho c^2) = \rho(1 + i/c^2). \tag{2.3}$$

We shall assume that ϵ is a given function of the variables p and ρ

$$\epsilon = \epsilon(p, \rho) . \tag{2.4}$$

This function will be different for different fluids. In classical theory we have

$$\epsilon = \frac{1}{\gamma - 1} p/\rho ,$$

where γ is the ratio of specific heats. The scalar p is called the pressure and the scalar ρ is called the rest density; ϵ is said to be the specific internal energy of the fluid and Eq. (2.4) is called the caloric equation of state of the fluid.

It follows from Eq. (1.4) that if x^μ is any vector field orthogonal to U^μ, then

$$T^\mu_\nu x^\nu = -p/c^2 x^\mu .$$

Thus T^μ_ν is a tensor which has one time-like proper vector with the proper value $\rho(1 + \epsilon/c^2)$ and three space-like proper vectors with the proper value $-p/c^2$.

These properties of T^μ_ν imply and are implied by the equation

$$\left(T^\mu_\nu + p/c^2 \delta^\mu_\nu\right) \left(T^\nu_\rho - (\sigma - p/c^2) \delta^\nu_\rho\right) = 0 ,$$

where

$$\sigma = \frac{1}{3} \sqrt{12S - 3T^2}$$

$$\frac{p}{c^2} = -\frac{T}{4} + \frac{1}{12} \sqrt{12S - 3T^2}$$

$$T = T^\mu_\mu$$

$$S = T^\mu_\rho T^\rho_\mu .$$

Thus the tensor T^μ_ν may be eliminated from the equations determining the $g_{\mu\nu}$, the Einstein field equations as was proposed by Rainich [1] and Misner and Wheeler [2] for the case of the electromagnetic field. Instead of following this proposal we shall see how in some cases we may reduce the field equations to a set of equations involving a single set of dependent variables. In order to do this we must discuss some consequences of the vanishing of the divergence of the tensor $T^{\mu\nu}$.

It should be remarked that a method of "solving" the field equations (1.1) is to choose the tensor $g_{\mu\nu}$ arbitrarily and then calculate the tensor $T^{\mu\nu}$. The resulting stress-energy tensor will not in general satisfy the algebraic conditions given in the preceding section. McVittie [3] has shown how the determination of the $T^{\mu\nu}$ from the $g_{\mu\nu}$ via the Einstein field equations may be used to solve various problems in classical hydrodynamics.

3. The Condition $T^{\mu\nu}_{;\nu} = 0$

Using Eq. (1.4) we may write this condition as

$$T^{\mu\nu}_{;\nu} = (\sigma U^{\nu})_{;\nu} U^{\mu} + \sigma U^{\nu} U^{\mu}_{;\nu} - \frac{p_{,\nu}}{c^2} g^{\mu\nu} = 0 . \qquad (3.1)$$

Multiplying by U_{μ} and summing we have

$$(\sigma U^{\nu})_{;\nu} = \frac{p_{,\nu}}{c^2} U^{\nu} . \qquad (3.2)$$

Hence

$$\sigma U^{\nu} U^{\mu}_{;\nu} = \frac{p_{,\nu}}{c^2} (g^{\mu\nu} - U^{\mu}U^{\nu}) . \qquad (3.3)$$

Equation (3.2) may be written as

$$(1 + i/c^2)(\rho U^{\nu})_{;\nu} + \frac{\rho T}{c^2} S_{;\nu} U^{\nu} = 0 \qquad (3.4)$$

where $S(p,\rho)$ and $T(p,\rho)$, the specific entropy and the temperature, are defined by the requirement that T be an integrating factor so that dS in the equation

$$T\, dS = d\epsilon + p d(1/\rho) = di - (1/\rho)\, dp$$

is a perfect differential.

Equation (3.3) is the relativistic formulation of the conservation of momentum equation. Equation (3.4) is a combination of two conservation equations of classical hydrodynamic the conservation of mass and the conservation of energy. Note that if the flow is isentropic, that is, if S = constant, then

$$(\rho U^{\nu})_{;\nu} = 0 \tag{3.5}$$

i.e., mass is conserved. We shall assume that in addition to the four equations (3.1), Eq. (3.5) holds. Then (3.4) reduces to

$$S_{,\nu}U^{\nu} = 0\ ; \tag{3.6}$$

that is, the entropy is constant along a streamline of the motion. This is the same situation as obtained in classical hydrodynamics where there are five conservation laws.

4. The Vorticity Vector

The vector defined by

$$v^{\mu} = \frac{1}{\sqrt{-g}}\,\epsilon^{\mu\nu\sigma\tau}\,U_{\sigma;\tau}\,U_{\nu} \tag{4.1}$$

will be called the vorticity vector. A flow will be said to be irrotational if

$$v^{\mu} = 0\ .$$

We define the vector

$$V_{\mu} = (1 + i/c^2)U_{\mu}\ . \tag{4.2}$$

Then

$$V_{\mu;\nu} = (1 + i/c^2)U_{\mu;\nu} + \frac{1}{c^2}\,i_{,\nu}U_{\mu}\ . \tag{4.3}$$

When we consider the specific enthalpy i as a function of pressure and entropy we may write

$$i_{,\nu} = \epsilon_{,\nu} + \frac{p_{,\nu}}{\rho} + p\left(\frac{1}{\rho}\right)_{,\nu}$$

$$i_{,\nu} = TS_{,\nu} + \frac{p_{,\nu}}{\rho}\ . \tag{4.4}$$

We may write

$$\Omega_{\mu\nu} = V_{\mu;\nu} - V_{\nu;\mu} = (1 + i/c^2)$$

$$\times \left[U_{\mu;\nu} - U_{\nu;\mu} + \frac{1}{c^2(1 + i/c^2)} (i_{,\nu} U_\mu - i_{,\mu} U_\nu) \right] . \tag{4.5}$$

It follows from Eq. (4.1) that

$$\sqrt{-g}\, \epsilon_{\mu\nu\sigma\tau} v^\mu = (U_{\sigma;\tau} - U_{\tau;\sigma}) U_\nu + (U_{\tau;\nu} - U_{\nu;\tau}) U_\sigma$$

$$+ (U_{\nu;\sigma} - U_{\sigma;\nu}) U_\tau .$$

Hence,

$$\sqrt{-g}\, \epsilon_{\mu\nu\sigma\tau} v^\mu U^\nu = (U_{\sigma;\tau} - U_{\tau;\sigma}) + U_{\tau;\nu} U^\nu U_\sigma - U_{\sigma;\nu} U^\nu U_\tau .$$

In view of Eqs. (3.3) this may be written as

$$\sqrt{-g}\, \epsilon_{\mu\nu\sigma\tau} v^\mu U^\nu = (U_{\sigma;\tau} - U_{\tau;\sigma}) + \frac{p_{,\rho}}{c^2 \rho (1 + i/c^2)}$$

$$\times \left[\left(\delta^\rho_\tau - U^\rho U_\tau \right) U_\sigma - \left(\delta^\rho_\sigma - U^\rho U_\sigma \right) U_\tau \right]$$

$$= (U_{\sigma;\tau} - U_{\tau;\sigma}) + \frac{p_{,\tau} U_\sigma - p_{,\sigma} U_\tau}{c^2 \rho (1 + i/c^2)} .$$

Thus

$$\Omega_{\mu\nu} - (1 + i/c^2) \sqrt{-g}\, \epsilon_{\sigma\tau\mu\nu} v^\sigma U^\tau = TS_{,\nu} U_\mu - TS_{,\mu} U_\nu ;$$

that is,

$$\Omega_{\mu\nu} = (1 + i/c^2) \sqrt{-g}\, v^\sigma U^\tau \epsilon_{\sigma\tau\mu\nu} + TS_{,\nu} U_\mu - TS_{,\mu} U_\nu . \tag{4.6}$$

Hence for isentropic irrotational flow we have

$$\Omega_{\mu\nu} = 0 . \tag{4.7}$$

As a consequence

$$U_\mu = \frac{1}{1 + i/c^2} \frac{\partial\theta}{\partial x^\mu}$$

where θ is an arbitrary scalar function. We may adapt our coordinate system so that $\theta = x^4$ and then

$$U_\mu = \frac{1}{(1 + i/c^2)} \delta^4_\mu \, . \tag{4.8}$$

It then follows from

$$U_\mu U^\mu = 1$$

that

$$g^{44} = (1 + i/c^2)^2 \, . \tag{4.9}$$

Hence for irrotational isentropic flow we may choose a co-moving coordinate system in which the stress energy tensor depends on g^{44} alone. Since in this case the enthalpy i is a function of pressure alone as is the rest density ρ; that is, Eq. (4.9) may be inverted to give p and ρ as functions of g^{44}. If these functions of g^{44} and Eq. (4.8) are used in the evaluation of the tensor $T^{\mu\nu}$, we may write the field equations (1.1) wholly in terms of the tensor $g_{\mu\nu}$. Thus the Einstein equations for the case of relativistic hydrodynamics involve a single set of dependent variables and one has a tractable problem.

Various exact and approximate solutions of these equations have been discussed. In particular, the problem of the one-dimensional motion of a gas analogous to the problem in classical hydrodynamics of the motion generated by pushing a piston into a tube filled with gas has been treated approximately. The procedure is to start with the special relativistic solution to the problem and to modify this solution in accordance with the Einstein field equations. It can be seen that the nonlinearities in this problem reside in the special relativistic approximations and that the gravitational corrections satisfy linear equations. Thus, in the zero-order approximation, shocks (discontinuities in the pressure, density, and tangents to the lines of the elements of the fluid) occur. It is unlikely that the gravitational corrections satisfying linear equations will remove them. We must therefore be prepared in general relativity to treat with and allow for coordinate transformations involving discontinuous derivatives. The metric tensor may thus become singular on certain hypersurfaces. Although such singularities are associated with coordinate transformations they are essential ones since the coordinates involved have an intrinsic and physical significance.

REFERENCES

[1] G. Y. Rainich, Electrodynamics in the general relativity theory. *Trans. Am. Math. Soc.* 27, 106-136 (1925).
[2] C. W. Misner and J. A. Wheeler, Geometrodynamics. *Ann. Phys. N.Y.* 2, 525-603 (1957).
[3] G. C. McVittie, "General Relativity and Cosmology," Chapters VI and VII. Wiley, New York, 1956.

GENERAL REFERENCES

A. H. Taub, A general relativistic variational principle for perfect fluids. *Phys. Rev.* 94, 1468-1470 (1954).
A. H. Taub, Isentropic hydrodynamics in plane symmetric space time. *Phys. Rev.* 103, 454-467 (1956).
A. H. Taub, Approximate solutions of the Einstein equations for isentropic motions of plane-symmetric distributions of perfect fluids. *Phys. Rev.* 107, 884-900 (1957).
A. H. Taub, Singular hypersurfaces in general relativity, III. *J. Math.* 1, 370-388 (1957).
A. H. Taub, On circulation in relativistic hydrodynamics. *Arch. Ratl. Mech. Anal.* 3, 312-324 (1959).

INVARIANCE GROUPS

Peter G. Bergmann

Syracuse University
Syracuse, New York

In the terminology of the physicist, an invariance group is a group of transformations which leave the form of some given set of laws of nature unchanged. This definition clearly leaves something to be desired, and I shall attempt to review the concept of invariance groups critically. Before I do so, permit me to give you the most important examples of "classical" invariance groups, that is to say, those whose importance at various stages of the history of physics is beyond question. I shall proceed from finite to transfinite invariance groups.

(1) Parity invariance indicates simply that our laws of nature are unchanged if we go over from a right-handed to a left-handed Cartesian coordinate system in describing arrangements in three-dimensional space. The "parity group" consists of but two elements, and its group table is adequately represented by the multiplication table of the two numbers 1 and -1. Five years ago it was discovered that there are particles in nature which have a built-in "chirality"; their spin and direction of forward motion form a screw that always has the same handedness. But as there exists a second type of particles with the reverse chirality (neutrinos, antineutrinos), many physicists now believe that the laws of nature will, after all, remain invariant under parity transformations provided that the purely geometric parity transformation is always combined with the interchange of these particles, with "Particle-antiparticle conjugation." At any rate, whether parity invariance is ultimately found to be valid in nature, it is a historically important example of a finite invariance group.

(2) The orthogonal group in three-dimensional space is a six-parameter Lie group, with respect to which the classical laws of physics are invariant. This invariance implies the absence of anisotropies and inhomogeneities in physical space *by themselves*. To the extent that such inhomogeneities are in fact found, they are believed to be the direct result of the presence of matter, in the widest possible sense, and thus part of the subject matter with which the laws of nature are to deal.

(3) If we extend the principle of homogeneity to the time axis and if we specify further that the (uniform) velocity of an "observer" has no effect on his findings concerning the laws of nature, then we expand the six-parameter orthogonal group into a ten-parametric group, either the Galilean group of Newtonian physics or the Lorentz group of special relativity. The formal difference between these two Lie groups is, of course, that they possess different multiplication tables; they are not isomorphic to each other. Historically, the Galilean group was constructed so that any time interval between two "events" in space-time is invariant under all space-and-time transformations of the group. In particular, as "action-at-a-distance" is a fundamental dynamical concept in Newtonian mechanics (the instantaneous exertion of forces between massive bodies over nonzero distances), the concept of simultaneity was to be absolute (invariant). The experiments with the electromagnetic field toward the end of the nineteenth century showed that the newly discovered laws of nature were not Galilean-invariant, but neither was it possible to ascertain from experiments the velocity of an observer, in particular the earth, with respect to some preordained frame of reference (the light ether). Under the Lorentz transformation group, and under the associated reformulation of the laws of nature, the "relativistic physics," the invariant interval is now a quadratic form that involves both spatial and time coordinate differences. This is not the place to attempt a complete discussion of the Minkowski universe and the Lorentz group, but I should like to mention one semiphilosophical point.

The relativistic laws have been confirmed primarily in microphysics, that is to say in experiments involving interactions between elementary particles, or at most between atoms or atomic nuclei. They have been confirmed not only for electromagnetic interactions but also for those known to physicists today as "strong" interactions and "weak" interactions. Both are types of nuclear forces; typical interactions of the former hold together the heavy components of atomic nuclei, whereas the latter are involved in transformations resulting in the emission of electrons and positrons (beta-decay) from atomic nuclei. On the other hand, our astronomical universe as a whole has properties that appear to justify a selection among all possible Lorentz frames. If we average the state of motion of matter over a sufficiently large portion of the universe, which includes several clusters of galaxies, then there emerges a prevalent state of motion, which is quite well defined, within a fraction of one percent of the speed of light (3×10^{10} cm sec^{-1}). Thus nature at large appears to favor a particular state of motion. To say that this preference applies only to one physical variable, the state of motion of large-scale matter, and hence does not represent a "law of nature," appears to me to involve one in a semantic ambiguity. Ordinarily, we speak of "laws of nature" as principles that apply to large numbers of separate physical systems and whose generality, accordingly, may be tested experimentally. In cosmological discussions, however, we are faced with a class of systems containing but one specimen, that is to say our universe. Cosmological principles thus apply at best to one example, but, by the same token, to all specimens of the class of objects (universes) available for observations. Thus, whether a cosmological "law" should be considered a "law of nature" or not is, perhaps, a question of semantics. It is, nevertheless, an important question, because the aim of cosmology is to study our cosmos as a physical object, to which the laws of nature apply.

As I pointed out, for better or for worse, of all the frames of reference that we might use to describe the universe, one set of space-time coordinate systems is clearly privileged, and it consists of all those frames in which the velocity of matter on the time axis (all space coordinates being zero) vanishes. Incidentally, it is impossible to construct a local velocity distribution of matter that is Lorentz-invariant, and which, therefore, would have an indeterminate average velocity. Because the Lorentz sphere of time-like directions is not compact, the mean kinetic energy density corresponding to such a distribution would be infinite.

(4) The "gage group" of electrodynamics is a continuous group that is infinite-dimensional. A gage transformation consists of the addition of an arbitrary gradient field to the four components of the potential field of the electromagnetic field. Thus any function of the four space-time coordinates

that satisfies reasonable continuity and differentiability conditions as well as reasonable boundary conditions at spatial infinity defines a gage transformation, and these transformations form an additive group. The gage transformations differ from all those transformations that I have mentioned previously in that knowledge of the potential field at all times $t \leq t_0$ does not preclude a gage transformation that is nontrivial for $t > t_0$. In view of the fact that the differential equations of the electromagnetic field, the Maxwell-Lorentz equations, are invariant under gage transformations, it follows that the past will determine the future, under these field equations, at best up to a gage transformation. As a matter of fact, knowledge of the gage-invariant quantities, the field intensities, at one time represents a satisfactory set of Cauchy-Kowalewsky data, and we may say that the Maxwell-Lorentz theory is deterministic with respect to all gage-invariant variables.

(5) The general theory of relativity is at present the most widely accepted theory of gravitation. Its invariance group is the group of curvilinear coordinate transformations, which may be narrowed down, depending on circumstances, by requirements of differentiability and by boundary conditions. Like the gage group, the group of coordinate transformations is infinite-dimensional. Its elements, the individual coordinate transformations, may be identified by the fixation of four arbitrary functions of the four space-time coordinates. The coordinate group (as I shall call it for short) is noncommutative. Whereas in electrodynamics a complete set of gage-invariant Cauchy data is well known, no simple, closed-form coordinate invariants have been constructed as yet in general relativity. On the other hand, the existence of such invariants, and the existence of the associated Cauchy data, has been demonstrated, by various methods. Some of the pioneering work is associated with the name of Lichnerowicz, and those of his students, foremost among them Fourès-Bruhat. Very recently new approaches have been originated by H. Bondi and his coworkers in London, by Roger Penrose, and, in this country, by R. Sachs. Much of this recent work is still in process of publication, but preprints have been widely circulated.

Physicists are interested in invariance groups for a number of reasons. The transformations belonging to the invariance group transform solutions of the dynamical laws into other solutions; solutions that can go over into each other by way of invariance transformations are considered to be physically equivalent. For instance, if we consider a solution of the equations of motion of celestial mechanics and then carry it over into another solution by means of an orthogonal transformation of the spatial Cartesian coordinates, then these two solutions of Newton's equations are considered to describe the same history of the same mechanical system. Thus the invariance group decomposes the set of all solutions of the dynamical laws into equivalence classes; the members of one equivalence class are supposed to represent "the same physical situation," whereas solutions belonging to different equivalence classes represent inequivalent, or intrinsically different, situations. Thus the invariance group plays a role in the physical interpretation of the solutions of the physical laws.

If the invariance group is a continuous group, then the infinitesimal group leads to constants of the motion because of Emmy Noether's theorems, at least if the dynamical laws can be stated in the form of a variational principle, a property that all known fundamental laws of physics possess. Noether's theorems state that the generator of an infinitesimal invariant transformation is a constant of the motion. For a field theory, Noether's theorems tell us that the generating vector field satisfies an equation of continuity. In other words, the invariance group of a theory is directly related to the existence of conserved quantities. The Galilean and Lorentz groups, for instance, give rise to the conservation of energy and linear momentum, of angular momentum, and of the reduced moment of the mass. Likewise, gage invariance is directly related to the conservation of electric charge. Conserved quantities, and conservation laws, are of special interest to the physicist, because the conservation of any quantity can usually be ascertained experimentally with greater assurance than, for instance, its rate of change if it is not conserved. Usually, invariance properties of a theory are in fact examined by way of the associated conservation laws. If, for instance, the energy of a mechanical system does not remain unchanged during a physical process, we conclude immediately that external forces must be involved, that the system is not "conservative."

Let me assure you, then, in summary of the foregoing, that invariance properties of a physical theory, or its symmetry properties, as they are often called, are of abiding concern to the practicing physicist, whether he be of the experimental or theoretical stripe. Nevertheless, the concept of the invariance group is open to a critique from two directions. We may either claim that the concept is too narrow, or that it is too wide.

In dealing with a specific theory we can construct many more constants of the motion than those associated with the conventional invariance group. Let me give you a specific example, vacuum electrodynamics, that is, electrodynamics in the absence of sources and in the absence of electric and magnetic materials. This theory can be analyzed very conveniently if we go over to the Fourier transforms (in three dimensions) of all of its field variables. Denoting the transforms of the scalar potential by ψ and of the vector potential by $\boldsymbol{\alpha}$, the Maxwell-Lorentz equations may be given the form

$$\boldsymbol{\beta} = i\mathbf{k} \times \boldsymbol{\alpha}\,, \qquad \boldsymbol{\epsilon} = -\frac{1}{c}\frac{\partial \boldsymbol{\alpha}}{\partial t} - i\mathbf{k}\psi$$

$$\mathbf{k} \cdot \boldsymbol{\epsilon} = 0\,, \qquad i\mathbf{k} \times \boldsymbol{\beta} - \frac{1}{c}\frac{\partial \boldsymbol{\epsilon}}{\partial t} = 0\,.$$

$\boldsymbol{\beta}$ is the Fourier transform of the magnetic field, $\boldsymbol{\epsilon}$ the Fourier transform of the electric field. Both are gage-invariant. It follows immediately from these equations that both the magnetic and the electric field Fourier transforms are perpendicular to the wave propagation vector $\mathbf{k}$; they are also perpendicular to each other. The integrals of these equations have the form

$$\boldsymbol{\alpha} = \boldsymbol{\alpha}_0 e^{-i\omega t} + i\mathbf{k}\lambda\,,$$

$$\omega = c|\mathbf{k}|\,.$$

$$\psi = \frac{c}{\omega}\mathbf{k} \cdot \boldsymbol{\alpha}_0 e^{-i\omega t} - \frac{1}{c}\dot{\lambda}\,,$$

In these expressions the vector $\boldsymbol{\alpha}_0$ is an arbitrary function of the argument $\mathbf{k}$ and constant in time, whereas λ is an arbitrary function of both $\mathbf{k}$ and the time coordinate t. This last scalar changes under gage transformations. The electric and magnetic field Fourier transforms take the form

$$\boldsymbol{\beta} = i\mathbf{k} \times \boldsymbol{\alpha}_0 e^{-i\omega t}\,, \qquad \boldsymbol{\epsilon} = -\frac{c}{\omega}\mathbf{k} \times \boldsymbol{\beta}\,;$$

that is to say, that portion of $\boldsymbol{\alpha}_0$ parallel to $\mathbf{k}$ contributes nothing to the electromagnetic field intensities, λ contributes nothing, and the "transverse" portion of $\boldsymbol{\alpha}_0$, the part of that vector field that is perpendicular to $\mathbf{k}$, determines the field entirely and for all time. That means that four real (or two complex) functions of $\mathbf{k}$ may be chosen arbitrarily, and these are precisely the constants of the motion of our physical model. To make matters entirely orthodox we must represent $\boldsymbol{\alpha}_0^T$ in terms of the electromagnetic field, but this is obviously possible. I shall not bore you with the arithmetical details; one must take into account both positive and negative frequencies; but one must also pay heed to reality requirements, and the field intensities determine, and they are in turn determined by, the stated number of constants of the motion.

Any functional of these constants of the motion generates an infinitesimal canonical transformation that maps every solution of the Maxwell-Lorentz equations on another solution. These mappings define a group, the group of invariant canonical transformations of this particular system of equations of motion; it is a subgroup of the group of canonical (though not necessarily invariant) transformations. Why not adopt them as *the* invariance group of our theory?

Intuitively, we might reply that the group of invariant transformations (in the technical sense in which this term is used in Hamiltonian theory) and the invariance group differ in that the invariance group is known on general physical principles before we integrate the equations of motion, whereas the group of invariant transformations becomes known to us only after we have succeeded in obtaining a complete set of integrals of the motion. Expressed differently, we might say that once we begin to couple one physical field to others, and thereby modify the variational principle on which the formal theory is based, the invariant transformations will change, whereas the structure of the invariance group does not. Unfortunately, such arguments are psychological rather than logical; that is to say, these arguments merely recast in formal language what we expect of the theory under development, not what we can deduce from the formal analysis of the finished product. A little later I shall adduce a formal point of view. But first I shall subject the concept of invariance group to yet another attack.

Let us consider once more electrodynamics. If we represent the theory in terms of electromagnetic potentials, we know that a solution remains a solution if we add to the potential four-vector field a four-gradient of an arbitrary scalar field; this change is what we call a gage transformation. As the quantities subject to physical observation are not the potentials but their skew-symmetric derivatives, the field intensities, and as the field intensities do not change at all under a gage transformation, we consider two potential vector fields that differ only by a gradient field as two descriptions of the same physical situation. At least in a singly connected domain, knowledge of the field intensities alone gives us complete information about the "intrinsic" properties of the electromagnetic field, sufficient, for instance, to obtain values of the electromagnetic field at some later (or earlier) time simply by integration of the field equations. Thus one might say that introducing the potentials, and with them the gage transformations, we create an entirely avoidable complication, and that the proper formulation of the theory can well dispense with both. From this point of view, the invariance group of electrodynamics, the gage group, is not a deep property of the theory but appears only as the result of our bungling. One could even assert that by yet more extensive proliferation of field variables one could extend the invariance group indefinitely.

This point of view appears entirely valid, except for the circumstance that the restriction to "intrinsic" variables, that is, to variables that are invariant under the invariance group of a theory, usually precludes our ability to derive the field equations from a variational principle. If we forego this requirement, then presumably any theory can be formulated entirely in terms of its intrinsic variables. Such a formulation will elucidate the actual degrees of freedom of the field, as well as the manner in which the history of the field is determined by the Cauchy data.

If all solutions of the dynamical laws that can be transformed into each other under the invariance group are considered physically equivalent, then we must conclude that no quantity can be determined experimentally whose value depends on the choice of frame. Only invariants can be determined, and that is why in general relativity the terms "invariant" and "observable" are now being used interchangeably. If we can elaborate a procedure by which a noninvariant variable may be measured, then our procedure itself is noninvariant, in that it includes some means of fixing the frame of description by physical apparatus; if so, this apparatus, consisting itself of material objects, is not subject to our invariance group, and that group is thereby shown to have less than universal validity.

Summarizing, then, my second attack on the concept of invariance group I should say that we may eliminate from the formulation of any theory all of its noninvariant variables, and with them the invariance group of that theory, admittedly sacrificing in the process our ability to state the whole theory in terms of a variational principle.

I do not believe that we have succeeded entirely in rescuing the concept of the invariance group from the double-barreled attack that I have outlined. Though not entirely persuasive, the strongest arguments for the defense appear to me to be the importance of universal symmetry principles in our grasp of the nature of things physical on the one hand, the importance of the principle of least action on the other. In what follows I shall direct your attention to a particular type of invariance group close to my heart, and that is the continuous invariance group involving arbitrary functions.

As I mentioned earlier, if the invariance group involves arbitrary functions of the space-time coordinates, then it is impossible, in principle, to predict the values of noninvariant quantities from any amount of Cauchy data. The two principally interesting invariance groups of this kind are the gage group of electrodynamics and the coordinate group of general relativity. In addition, theoretical physicists have constructed similar groups, partly by way of speculation, partly in order to have available practice examples, which are called Yang-Mills groups, after their originators. I shall call all such groups function groups, a terminology consistent with that introduced by S. Lie.

The chief difference between a function group as invariance group and a Lie group, as far as observability is concerned, is this: In a Lie group the frame of reference is determined, incidentally, by the Cauchy data. Thus, even if a certain quantity is not an invariant, it assumes an invariant significance in connection with information given in a different space-time domain. Let us, for instance, consider a system consisting of a single free particle. The only intrinsic property of a classical free particle is its mass. In the absence of specific information concerning the chosen frame of reference a statement of the relativistic energy of the particle conveys no intrinsic information; but if we state, for instance, the present location of the particle, as well as its location at a specified earlier time, in addition to its relativistic energy, then the totality of this information enables us to determine the rest mass of the particle, because with the information at hand we can transform to a preferred frame of reference, the rest frame of the particle.

By contrast, in the presence of a function group any information that we may possess concerning the frame of reference applies only to the domain in space-time for which we are given these data; noninvariant information applying to a different and disjoint domain cannot be correlated with the original information to give us intrinsic information that is not already available in each domain separately. In other words, if in general relativity I start with some coordinate system at one time, there is nothing in the dynamical laws to restrict the coordinate system at some subsequent time. Any noncovariant quantities given to me at some later time will be of intrinsic value only to the same extent as they would be if I possessed no initial-value information. How, then, can I hope to obtain intrinsic information about a physical situation at all?

There are several answers to this question. One possible answer is provided by the experience of electrodynamics. There may exist a simple algorithm that enables me to distill, out of the noninvariant quantities, complete invariant information, in the form of differential invariants or otherwise. In general relativity the situation is not that simple. However, Géhéniau, Debever, and Komar have developed the technique of "intrinsic coordinates," which leads to the desired goal, though perhaps in a clumsy fashion. This technique may be described either formally or intuitively. Formally, the intrinsic coordinates are simply coordinates which are uniquely determined by intrinsic local properties of the field. For instance, the metric field in a pseudo-Riemannian manifold that obeys Einstein's equations is characterized in part by four scalars that are obtained algebraically from the components of Weyl's tensor (also known as the conformal curvature tensor). These four scalars may be adopted as the "intrinsic" coordinates, and that is a fixation of coordinates by means of purely formal properties of the geometry that permits no further coordinate transformations. Intuitively, we may say that intrinsic properties of the metric field may be obtained if we furnish the values of at least five scalar fields at the same four-dimensional point in space-time. We need at least five such fields, because the first four are required merely to fix the point at which we make our determinations; beginning with the fifth piece of data we get information that is characteristic of particular Riemann-Einstein manifolds. It has been shown that in order to fix a Riemann-Einstein manifold completely we must obtain four pieces of such data on each point of a suitably chosen three-dimensional domain.

I shall now return to Noether's theorems and their bearing on a function group. On the one hand, these theorems tell us that the generator of an invariant transformation is a constant of the motion; on the other hand, by assumption our invariant transformations involve arbitrary functions of the space-time coordinates, which, of course, must also appear as coefficients in the generator. It follows that the generating functions must vanish. And thus we are confronted with the necessity that in the presence of a functional invariance group there must exist vanishing functions, or functionals, of the canonical field variables equal in number to the number of arbitrary functions of the time entering into the invariance group. These vanishing functions of the canonical variables have been called constraints, because they establish relationships between the canonical variables free of time derivatives. Moreover, because these constraints are generators of transformations belonging to the invariance group, the commutator between any two of these constraints, their Poisson bracket, must be a constraint again. In the terminology originated by Dirac, the generators of the invariance group must be first-class constraints.

In retrospect, this result may be converted into a definition. We may call an invariance group that group of transformations of variables generated by the first-class constraints. For this definition to be precise, we should include the Hamiltonian of the theory among the first-class constraints, something that is automatically the case in general relativity but may have to be achieved by a formal trick in other theories. Any other invariant transformations will be generated by constants of the motion which do not vanish: in this manner the invariance subgroup is identified formally within the group of invariant transformations. To my knowledge, this is the only formal definition that has been discovered as yet. Needless to say, one can free this definition from the context of the Hamiltonian formalism, if desired. The concept of invariant canonical transformations may be introduced into the variational principle directly; the generators belonging to the invariance group will again vanish if the field equations are satisfied.

Before closing, I should like to discuss briefly some recent work which we did on the invariance group of general relativity. This group, as I mentioned earlier, is the group of curvilinear coordinate transformations; the four arbitrary functions involved are the new coordinates of a four-dimensional point as functions of its original coordinate values. The corresponding infinitesimal group may be thought of as corresponding to the totality of all four-dimensional contravariant vector fields, the field of infinitesimal displacements. Stated in these simple terms, the definition is, of course, incomplete. I have said nothing as yet concerning the continuity and differentiability requirements on the functions involved. These requirements depend, to some extent, on the applications that one has in mind. As we are generally interested only in the domain of real coordinate values we should not restrict ourselves to analytic functions. Usually one requires that there should exist n continuous derivatives, and that the Jacobian should be nonzero. These requirements frequently apply to a four-dimensional domain that is topologically equivalent to less than the Minkowski universe; hence the invariance group in question is perhaps more correctly described as a pseudogroup. The exclusion of some regions from the requirements of continuity, differentiability, and reversibility is based on the circumstance that, for instance, the location of a particle represents a singularity of the field, and thus there is no reason why the coordinate transformations themselves might not be singular there as well.

For many physical situations the pseudo-Riemannian manifold approaches the metric properties of a Minkowski universe at spatial infinity, and it is thus desirable to use coordinate systems in which the metric tensor components approach the Minkowski-Lorentz values asymptotically. If we restrict our consideration to these manifolds, then the "natural" invariance group is that set of curvilinear coordinate transformations that maintain the asymptotic form of the metric tensor. This group, or pseudogroup, of transformations will be characterized more precisely in terms of the specific asymptotic properties we wish to maintain. Several of these asymptotically restricted groups have been investigated, by Bondi and his coworkers, by Sachs, and by myself, in part jointly with Robinson and Schücking. One question of particular interest is this: If the metric is to be, and to remain, asymptotically Lorentzian, is it perhaps possible to construct variables, preferably a complete set of them, which are Lorentz covariants, rather than general coordinate covariants? In other words, is it perhaps possible to segregate in some fashion the remaining arbitrariness of the coordinate transformations in the interior from the asymptotic Lorentz characteristics at spatial infinity, and thus to reduce the general theory of relativity to the form of a Lorentz-covariant field theory? This question has become even more interesting because several workers, such as Fock in the Soviet Union, and Arnowitt, Deser, and Misner in this country, have actually proposed methods for accomplishing this aim.

Whether these proposals are in fact viable is not easy to decide. I should like to indicate two possibilities how the reduction to a Lorentz-covariant theory might be accomplished. One is that the set of coordinate systems admissible is restricted by requirements in addition to the asymptotic behavior of the metric, to a ten-parametric set, and that then a group of transformations is constructed which maps this set on itself. Fock has pronounced a conjecture to the effect that harmonic coordinates, along with suitable asymptotic restrictions, form such a ten-parametric set. He has not indicated how he proposes to construct the corresponding transformation group: but without one the transformation laws of the quantities forming the dynamical variables of the theory hang in midair. As far as this approach is concerned, I consider the question of existence unsettled.

The other approach accepts the existence of the group of curvilinear transformations further circumscribed by asymptotic requirements. It asks whether there exists a homomorphism between that group and the Lorentz group. This investigation has been completed for two groups that are defined by sufficiently different asymptotic requirements that they are almost certainly not isomorphic to each other. However, surprisingly perhaps, the results of the group-theoretical investigations on both are strikingly similar: In both cases there exists a homomorphism with the homogeneous Lorentz group, but not with the inhomogeneous Lorentz group. It is, of course, possible that the analysis of yet other transformation groups will lead to different answers. But as far as present results go, I think that they ought to be given considerable credence. It appears that even with asymptotic boundary conditions imposed general relativity does not permit the construction of Lorentz-covariant quantities unless they are also constants of the motion. It would thus appear that the "softening up" of the original Minkowski geometry by the introduction of pseudo-Riemannian metric fields is not a reversible step, and that the new space-time concepts of general relativity are with us for good.

Perhaps the situation in general relativity, in which a fairly specific and also quite important problem is now being attacked with more sophisticated tools than were thought necessary a few decades ago reflects the general status of invariance groups in physics; fundamental, mystifying, challenging.

THE CAUCHY PROBLEM AND ENTROPY IN CHARGED, COMPRESSIBLE RELATIVISTIC SELF-INDUCTIVE FLUIDS *

N. Coburn

University of Michigan
Ann Arbor, Michigan

Part I

1. Introduction

We have divided this paper into two parts. In Part I, we shall study the Cauchy problem for three hyperbolic systems of partial differential equations: (1) the two-dimensional [(x, t), where x is the space variable and t is the time] wave equation; (2) a hyperbolic system of quasi-linear partial differential equations which applies to potential flow of a compressible non-charged Newtonian fluid in n independent variables (where $n = 2$, 3, or 4); (3) a hyperbolic system of partial differential equations which corresponds to the flow of a fictional (and simple) compressible charged relativistic fluid. By use of the first example, we shall show that for properly given Cauchy data along the line $t = 0$, $0 \leq x \leq b$, a *unique analytic solution* can be determined in the domain of dependence of the initial data [1] which is bounded by $t = 0$, $x - t = 0$, $x + t = b$. Further, we shall show that along the line $x - t = 0$, such Cauchy data *do not determine a unique* analytic solution. Hereafter,

*This work was done under National Science Foundation Grant G-10103 and Air Force Office of Scientific Research Grant 20-63, administered by the Office of Research Administration of the University of Michigan.

our main interest will be in determining the $(n - 1)$-dimensional submanifolds, S_{n-1}, of a given n-dimensional space, V_n, along which the Cauchy problem has no unique solution. Such S_{n-1} are possible characteristic wave fronts. In fact, the S_{n-1} for which the Cauchy problem has no unique solution are exactly those S_{n-1} where the normal derivatives of some of the dependent variables may be discontinuous ([2], Section 4). The second example will indicate some of the *essential steps in the procedure for determining such* S_{n-1}. The general procedure is discussed in [2]. In the third and final example, we shall show that the *necessary and sufficient condition for the coupling of mechanical and electromagnetic* phenomena in the determination of the above S_{n-1} (actually, S_3 in the case of relativistic fluids) is that the *current vector* J^k $(k = 1, 2, 3, 4)$ *cannot be specified on* S_3 *by the Cauchy data.* This last statement means that J^k *is discontinuous as one crosses a characteristic wave front or* $[J^k]$, the jump of J^k, is not zero [2].

In Part II of this paper, we shall limit our study to the determination of the modifications in the basic Cauchy equations ([2], Eqs. (3.13), (3.14), (3.23), (3.24)) of the noninductive case for the case when self-induction is present. It should be noted that the *self-induction terms are defined to be those which are multiplied by the coefficient* $(1 - \lambda\mu)$, where λ is the dielectric constant and μ is the constant magnetic permeability. Quan [3] has shown how these self-inductive terms enter into the Maxwell equations. Since these terms also enter into the entropy relation, we neglected them in our previous work ([2], Section 10). However, in Part II of this paper, we shall determine: (1) how these terms effect the entropy relation; (2) and then how these terms effect the basic Cauchy equations.

Before completing this introduction, we shall indicate the relation between our work on the Cauchy problem (and the corresponding problem of determining characteristic wave fronts and their speeds of propagation) for charged compressible relativistic fluids and the related work of other authors. A general discussion of the Cauchy problem for relativistic fluids was given by Lichnerowicz [4] in 1955. However, he neglected self-induction effects in charged fluids and studied the case where the current J^k, is known on the given S_3 (or is continuous on a characteristic S_3). This leads to the case where the coupling of electromagnetic and mechanical effects is not present. Synge [5a] has recently given an interesting but similar treatment of relativistic fluids. In some recent work, Abraham [6] extended Quan's theory [3] to the case where λ, μ (the dielectric coefficient and the coefficient of magnetic permeability) are variable in space-time. Both Abraham and Quan assume that the current is known along the given S_3. Further, by use of the theory of weak shocks ([2], Section 8), Foures-Bruhat [7] has shown that coupling of mechanical and electromagnetic effects exists in relativistic non-inductive fluids. Finally, Saini [8] has used a discontinuity theory approach ([2], Section 8) to determine characteristic S_3 for charged compressible relativistic self-inductive fluids of infinite conductivity where coupling of electromagnetic and mechanical effects exist.

2. Example 1; Solutions of the Wave Equation ($u_{tt} = u_{xx}$, where $u_{xx} \equiv \partial^2 u/\partial x^2$, etc.)

We assume that the boundary conditions are given by the Cauchy data

$$u(x,0) = f(x), \qquad u_t(x,0) = g(x) \tag{2.1}$$

where $f(x)$ and $g(x)$ are of class C^∞. From (2.1), the assumption of analyticity of the solution of the partial differential equation (the wave equation), and from the wave equation itself, we find by differentiation

$$u_{tt}(x,0) = f''(x), \qquad u_{ttt}(x,0) = u_{txx}(x,0) = g''(x), \tag{2.2}$$

where f'', g'' denote the second derivatives of f, g, respectively. By use of a similar procedure, we can determine the normal derivatives of higher order. Thus, we find for the fourth normal derivative

$$u_{tttt}(x,0) = u_{ttxx}(x,0) = u_{xxxx}(x,0) = f^{\mathrm{IV}}(x) \tag{2.3}$$

and for the general $(2n)$th normal derivative and $(2n+1)$th normal derivative

$$\left.\frac{\partial^{2n} u}{\partial t^{2n}}\right)_{x,0} = f^{(2n)}(x), \qquad \left.\frac{\partial^{2n+1} u}{\partial t^{2n+1}}\right)_{x,0} = g^{(2n)}(x) . \tag{2.4}$$

Thus, the desired analytic solution of the wave equation is

$$u(x,t) = f(x) + tg(x) + \frac{t^2}{\underline{/2}} f''(x) + \frac{t^3}{\underline{/3}} g''(x) + \dots$$

$$\frac{t^{2n}}{\underline{/2n}} f^{(2n)}(x) + \frac{t^{2n+1}}{\underline{/2n+1}} g^{(2n)}(x) + \dots \tag{2.5}$$

Now, we shall show that two families of curves exist along which Cauchy data are insufficient to determine a unique solution of the wave equation by use of the above methods. Consider the change of independent variables defined by

$$\xi = x - t, \qquad \eta = x + t. \tag{2.6}$$

The curves ξ = constant, η = constant are called *characteristic curves.* In terms of the variables ξ, η, the wave equation becomes

$$u_{\xi\eta} = 0. \tag{2.7}$$

It can be easily verified that the Cauchy data

$$u(\xi,0) = h(\xi), \qquad u_\eta(\xi,0) = k(\xi) \tag{2.8}$$

are insufficient to obtain a unique solution of (2.7) by use of our previous method. Further, it can be shown [9] that no unique solution (not necessarily analytic) exists for the equation (2.7) for Cauchy data of type (2.8). Essentially, this is due to the fact that the *second derivative of* u *along the normal to* $\eta = 0$ *cannot be determined.*

3. Example 2; The Characteristic Manifolds of the Quasi-Linear Second Order Hyperbolic Equation

Here, we consider the partial differential equation

$$a^{jk}\partial_j\partial_k\Omega = 0, \qquad a^{jk} = a^{jk}(x^p, \Omega, \partial_s\Omega), \qquad j, k, \dots = 1, 2, \dots n \tag{3.1}$$

where ∂_j denotes $\partial/\partial x^j$. If we introduce the new dependent variables

$$v_k = \partial_k\Omega \tag{3.2}$$

then the single equation (3.1) may be replaced by the equivalent system

$$a^{jk}\partial_j v_k = 0, \qquad \partial_j v_k - \partial_k v_j = 0. \tag{3.3}$$

Note, in (3.1), (3.3), the Einstein summation convention of summing on a repeated upper and lower index has been used. The following Cauchy data will be considered to be known: (1) the hypersurface S_{n-1}: $\phi(x^j) = 0$, and hence $(n-1)$ linearly independent tangent vectors $\underset{a}{t}^j$ $(a = 1, 2, \ldots n-1)$ at each point P of S_{n-1} and the normal vector n_j or $\partial_j\phi$ of S_{n-1} at P; (2) Ω, v_j, $\underset{a}{t}^k\partial_k v_j$ on S_{n-1}. If we decompose $\partial_j v_k$, we find

$$\partial_j v_k = n_j U_k + \sum_a \underset{a}{t}_j \underset{a}{V}_k . \tag{3.4}$$

Forming the scalar products of (3.4) with $\underset{a}{t}^j$, we see that $\underset{a}{V}_k$ are known multiples of the known tangential derivatives of v_j. However, U_k is a known multiple of the *unknown normal derivative of* v_j. In fact, a *sufficient condition for the non-uniqueness* of an analytic solution [see (2.5) and Example 1] for the system (3.3) *is that* $n^j n^k \partial_j v_k$ *cannot be determined from the given Cauchy data.*

Our problem is to formulate this last condition in analytic form. If we substitute (3.4) into (3.3), we obtain

$$a^{jk} n_j U_k = Q \tag{3.5}$$

$$n_j U_k - n_k U_j = W_{jk} \tag{3.6}$$

where Q, W_{jk} are known on the given S_{n-1}. The components of U_k which lie along $\underset{a}{t}_k$ can be determined from (3.6); the component of U_k which lies along n_k is a known multiple of $n^j n^k \partial_j v_k$. This normal component of U_k does not occur in (3.6). Further, *this component of* U_k *will not be present in (3.5) if and only if*

$$a^{jk} n_j n_k = 0 . \tag{3.7}$$

Hence, if the given S_{n-1} is determined by (3.7) then *Cauchy data is insufficient to determine a unique analytic solution of (3.1).*

4. Example 3; The Cauchy Problem for a Fictional Charged Compressible Relativistic Fluid

We consider a fluid with fictional (but simple) mechanical stress energy tensor

$$M^{jk} = \rho u^j u^k - p g^{jk}, \qquad j, = k = 1, 2, 3, 4 \tag{4.1}$$

where p, ρ are the pressure and density, respectively, g^{jk} is the metric tensor, u^j is the unit time-like vector which is tangent to a world-line in the curvilinear coordinate system x^j. The fluid will be assumed to be isentropic so that p and a are determined by

$$p = p(\rho), \qquad a^2 = \frac{dp}{d\rho}\,. \tag{4.2}$$

Hence, the divergence of (4.1) reduces to (where ∇_j denotes the covariant derivative)

$$\nabla_j M^{jk} = (u^j u^k - a^2 g_{jk}) \nabla_{jp} + \rho(u^j \nabla_j u^k + u^k \nabla_j u^j)\,. \tag{4.3}$$

The fictional (but simple) fluid is assumed to have the symmetric electromagnetic stress energy tensor

$$E^{jk} = \frac{g^{jk}}{4} F_{pq} F^{pq} + g^{kp} F^{jq} F_{qp} \tag{4.4}$$

where F^{jk} is the skew-symmetric electromagnetic field tensor which satisfies the Maxwell relations

$$\nabla_j F^{jk} = J^k \tag{4.5}$$

$$\nabla_j F_{pq} + \nabla_q F_{jp} + \nabla_p F_{qj} = 0\,. \tag{4.6}$$

The vector J^k in (4.5) is the current vector. Further, from (4.4), (4.5), it can be shown that (cf. [4], p. 47)

$$\nabla_j E^{jk} = F^{pk} J_p\,. \tag{4.7}$$

If we assume that the *divergence of the total stress energy tensor,* T^{jk}, *vanishes* where

$$T^{jk} = M^{jk} + E^{jk} \tag{4.8}$$

then from (4.3), (4.7), we find

$$(u^j u^k - a^2 g^{jk}) \nabla_j \rho + \rho(u^j \nabla_j u^k + u^k \nabla_j u^j) + F^{pk} J_p = 0. \tag{4.9}$$

Finally, by differentiating the relation

$$u^k u_k = 1 \tag{4.10}$$

we find

$$u_k \nabla_j u^k = 0 . \tag{4.11}$$

The relations (4.5), (4.6), (4.9), (4.11) constitute the basic system for our study of the *characteristic manifolds* of the charged compressible relativistic fictional fluid.

In terms of any three known independent vectors, $\underset{a}{t}_j$, $a = 1, 2, 3$, which lie in the local tangent hyperplane at any point P of the known hypersurface, $\phi(x^j) = 0$, the known normal vector $\phi_j \equiv \partial\phi/\partial x^j$, we can decompose $\nabla_j u_k$, $\nabla_j F_{pq}$, $\nabla_j \rho$ into the *known* $\underset{a}{A}_k$, $\underset{a}{C}_{pq}$, $\underset{a}{D}$ (determined by the Cauchy data), and the *unknown* U_k, G_{pq}, P by means of

$$\nabla_j u_k = \phi_j U_k + \sum_a \underset{a}{t}_j \underset{a}{A}_k \tag{4.12}$$

$$\nabla_j F_{pq} = \phi_j G_{pq} + \sum_a \underset{a}{t}_j \underset{a}{C}_{pq} \tag{4.13}$$

$$\nabla_j \rho = \phi_j P + \sum_a \underset{a}{t}_j \underset{a}{D}. \tag{4.14}$$

By substituting (4.12) into (4.11), we obtain

$$u_k U^k = 0 \tag{4.15}$$

plus some conditions on the known $\underset{a}{A}_k$. Since F_{pq} is skew-symmetric, it follows that G_{pq} of (4.13) is skew-symmetric. By decomposing G_{pq}, we see that a tangent vector T^j exists such that

$$G_{pq} = \phi_p T_q - \phi_q T_p + \sum_{a>b} \underset{ab}{u}\, \underset{a}{t_p}\, \underset{b}{t_q} . \tag{4.16}$$

The scalars $\underset{ab}{u}$ are known [determined by (4.6) and the Cauchy data]. If we substitute (4.16) into (4.13) and then substitute the resulting relation into the Maxwell equations (4.5), we find that [2]

$$J^k = \Phi T^k + \sum_a \underset{a}{t_j}\, \underset{a}{C^{jk}}, \qquad \Phi = g^{jk} \phi_j \phi_k \tag{4.17}$$

Note, J^k is *unknown* since T^k is considered to be *unknown.* Finally, if we substitute (4.12), (4.14), (4.17) into (4.9) and if we use the notation

$$L \equiv u^j \phi_j, \qquad \phi^j \equiv g^{jk} \phi_k \tag{4.18}$$

then (4.9) becomes

$$(Lu^k - a^2 \phi^k)P + \rho(LU^k + u^k \phi_j U^j) + \Phi F^{pk} T_p = R^k \tag{4.19}$$

where R^k is known. Thus, the basic system of equations consists of (4.15), (4.19) and the relation

$$\phi_j T^j = 0 . \tag{4.20}$$

The equations (4.15), (4.19), (4.20) form an *undetermined system of six linear algebraic equations in nine unknowns* P, U^j, T^j. In our previous paper [2], we showed that by assuming infinite conductivity ($F^{jk} u_k = 0$), the system becomes determined. Further, we generalized the procedure of example 2 for determining those S_3 for which the Cauchy problem has no unique solution. Again, we note that if J^k is *known* then T^k is *known* and the underdetermined system (4.15), (4.19), (4.20) reduces to a determined system consisting of (4.15), (4.19) but the *manifolds* for which the Cauchy problem has no unique solution *(essentially, determined by equating the determinant of the unknowns* P, U^k *to zero) are independent of the electromagnetic field.*

Part II

1. Introduction

In this part of the paper, two results of a previous paper [2] are generalized. First, *we obtain the general relation for the rate of change of entropy along a world-line (streamline)* in terms of the electromagnetic field variables for self-inductive (and noninductive) charged, compressible, nonviscous relativistic fluids. This relation reduces to that obtained by Taub for the noninductive case (cf. [2], Eq. (2.18), p. 364) and to that obtained by Saini for the self-inductive, infinite conductive case [8]. Secondly, we generalize to the self-inductive case the expression for the current density vector (cf. [2], pp. 367 and 369) in terms of a vector tangent to the initial hypersurface for the Cauchy problem (or to the characteristic hypersurface for discontinuity theory). Again, we show that *the unknown component of current is tangent to the initial hypersurface.* The Cauchy equation stemming from the conservation of total stress energy *differs from its non-inductive counterpart mainly in the occurrence of three new terms. The other Cauchy equations are unaltered.*

2. The Basic Relations

Let x^j, $j = 1, 2, 3, 4$ denote a curvilinear coordinate system in V_4, the four-space of general relativity. The element of arc, ds, will be determined by

$$\sigma\, ds^2 = g_{jk}\, dx^j\, dx^k \tag{2.1}$$

where σ is $(+1)$ for time-like displacements and (-1) for space-like displacements and g_{jk} is the metric tensor. Further, let ∇_j denote the covariant derivative and let u^j, ρ, S denote the unit time-like vector tangent to a world-line (streamline), the rest density, the specific entropy, respectively. The *conservation of mass* law [10] may be expressed by

$$\nabla_j(\rho u^j) = 0\,. \tag{2.2}$$

The symmetric stress energy tensor T_{jk} can be decomposed into two symmetric tensors M_{jk}, E_{jk}

$$T_{jk} = M_{jk} + E_{jk} . \tag{2.3}$$

The symmetric mechanical stress energy tensor M_{jk} is given by

$$M_{jk} = c^2 \eta u_j u_k - p g_{jk} \tag{2.4}$$

where p is the pressure which is a function of the density, ρ, and the entropy, S. Further by definition [10], the variable η is

$$\eta = \rho \left(1 + \frac{e}{c^2} + \frac{p}{\rho c^2}\right) \tag{2.5}$$

where e is the internal rest energy. We will assume in our future work that η and S are the two independent thermodynamical variables and hence

$$p = p(\eta, S), \qquad e = e(\eta, S), \qquad \rho = \rho(\eta, S). \tag{2.6}$$

For self-inductive charged fluids, the symmetric electromagnetic stress energy tensor E_{jk} is given by [3]

$$E_{jk} = \tau_{jk} - (1 - \lambda\mu)\tau_{jp} u^p u_k \tag{2.7}$$

where λ and μ are the dielectric capacity and magnetic permeability, respectively. The Maxwell stress tensor τ_{jk} can be expressed in terms of the skew-symmetric electromagnetic field tensors F_{jk}, H_{jk} by

$$\tau_{jk} = F^{pq} H_{pq} \frac{g_{jk}}{4} - F_{pj} H^{p}_{\cdot k} \tag{2.8}$$

where F_{jk}, H_{jk} are related by

$$\mu F_{jk} = H_{jk} + (1 - \lambda\mu)(H_{pj} u^p u_k - H_{pk} u^p u_j). \tag{2.9}$$

The terms which contain the multiplier $(1 - \lambda\mu)$ are the self-induction terms. We note *for both the case of non-inducation and the case of self-induction and infinite conductivity,* ($H^{jp}u_p = 0$), *(2.9) reduces to*

$$\mu F_{jk} = H_{jk} \,. \tag{2.9a}$$

Also, we see that for infinite conductivity, the *self-induction case leads to only one term more in* E_{jk} *than the non-inductive case*; that is, by substituting (2.9a) into (2.8) and then using (2.8) in (2.7), we obtain

$$\mu E_{jk} = H^{pq} H_{pq} \frac{g_{jk}}{4} - H_{pj} H^{p}_{\cdot k} - \frac{(1-\lambda\mu)}{4} H^{pq} H_{pq} u_j u_k \,. \tag{2.9b}$$

Finally, we note that the conservation law of total stress energy is

$$\nabla_j (M^{jk} + E^{jk}) = 0 \,. \tag{2.10}$$

Now, we shall consider the Maxwell relations and the stress energy conservation law. By differentiating (2.4), we obtain

$$\nabla_j M^{jk} = c^2(\eta u^j \nabla_j u^k + \eta u^k \nabla_j u^j + u^j u^k \nabla_j \eta) - g^{jk} \nabla_j p \,. \tag{2.11}$$

Again, by differentiation of (2.8), we find

$$\nabla_j \tau^{tk} = \frac{g^{tk}}{4} (F^{pq} \nabla_j H_{pq} + H_{pq} \nabla_j F^{pq})$$

$$- H_p^{\cdot k} \nabla_j F^{pt} - F^{pt} \nabla_j H_p^{\cdot k} \,. \tag{2.12}$$

In order to simplify the conservation law (2.10), we shall need the Maxwell relations. If J^k is the current four-vector, then these relations are

$$\nabla_j H_{pq} + \nabla_q H_{jp} + \nabla_p H_{qj} = 0 \tag{2.13}$$

$$\nabla_j F^{jk} = J^k \,. \tag{2.14}$$

First, we note that (2.13) leads to [5b]

$$F^{pq}(\nabla_j H_{pq} - 2\nabla_q H_{pj}) = 0 \,. \tag{2.15}$$

If we replace the index t by j in (2.12), use (2.14), (2.15), then (2.12) becomes

$$\nabla_j \tau^{jk} = \frac{g^{jk}}{4} (H_{pq} \nabla_j F^{pq} - F_{pq} \nabla_j H^{pq}) + H^{pk} J_p \,. \tag{2.16}$$

By use of a rest frame of reference at any point P (i.e., one such that $u^\alpha = 0$, $\alpha = 1, 2, 3$; $u^4 = 1$), Synge [5b], p. 417) has shown that the first term on the right-hand side of (2.16) vanishes. Thus, (2.16) becomes

$$\nabla_j \tau^{jk} = H^{pk} J_p \,. \tag{2.17}$$

Then, by use of (2.17), the divergence of (2.7) leads to

$$\begin{aligned} \nabla_j E^{jk} = H^{pk} J_p - (1 - \lambda\mu) \Big[&- H^{qp} J_p u^k u_q \\ &+ \left(F^{tq} H_{tq} \frac{g^{jp}}{4} - F^{tj} H_t^{\cdot p} \right) u^k \nabla_j u_p \\ &+ \left(F^{tq} H_{tq} \frac{u^j}{4} + F^{qj} H_{pq} u^p \right) \nabla_j u^k \Big] \,. \end{aligned} \tag{2.18}$$

3. Entropy in Charged (Self-Inductive) Compressible Relativistic Fluids

We assume that the first law of thermodynamics is valid for arbitrary time-like displacements and hence for all displacements. If T is the absolute temperature, then this law is

$$\nabla_j e = T \nabla_j S + \frac{p}{\rho^2} \nabla_j \rho \,. \tag{3.1}$$

Again, we note that u^j is a time-like unit vector and hence

$$u^j u_j = 1, \qquad u^j \nabla_k u_j = 0. \tag{3.2}$$

Further, we introduce the relativistic electric field by the definition

$$E^k \equiv H^{pk} u_p \,. \tag{3.3}$$

Substituting (2.11), (2.18) and (3.3) into (2.10), we obtain after multiplying by u_k

$$c^2(\eta\nabla_j u^j + u^j \nabla_j \eta) - u^j \nabla_j p - E^p J_p - (1-\lambda\mu)\left(-E^p J_p + \frac{F^{tq}}{4} H_{tq} \nabla_j u^j - F^{tj} H_t^{\cdot p} \nabla_j u_p\right) = 0. \tag{3.4}$$

Differentiating the relation (2.5) for η, and then forming the scalar product with u^j, we find after eliminating, $u^j \nabla_j e$, by use of (3.1)

$$c^2 u^j \nabla_j \eta = c^2 \frac{\eta}{\rho} u^j \nabla_j \rho + \rho T u^j \nabla_j S + u^j \nabla_j p. \tag{3.5}$$

If we eliminate $u^j \nabla_j \rho$ in the right-hand side of (3.5) through use of the conservation of mass relation (2.2), and then substitute (3.5) into (3.4), we obtain

$$T u^j \nabla_j S = E^p J_p + (1-\lambda\mu)\left(-E^p J_p + \frac{F^{tq}}{4} H_{tq} \nabla_j u^j - F^{tj} H_t^{\cdot p} \nabla_j u_p\right). \tag{3.6}$$

From (3.6), we see that the *flow is "adiabatic"* (that is, no heat is conducted by fluid particles or entropy is constant along a streamline) when: (1) the fluid is noninductive and

$$E^p J_p = 0 \tag{3.7}$$

or; (2) *the fluid is self-inductive and*

$$\lambda\mu E^p J_p + (1-\lambda\mu)\left(\frac{F^{tq}}{4} H_{tq} \nabla_j u^j - F^{tj} H_t^{\cdot p} \nabla_j u_p\right) = 0 \tag{3.8}$$

or; (3) *the fluid is infinitely conductive* $(E_p = 0)$ *self-inductive*, and

$$\frac{F^{tq}}{4} H_{tq} \nabla_j u^j - F^{tj} H_t^{\cdot p} \nabla_j u_p = 0. \tag{3.9}$$

The result (3.7) is due to Taub (cf. [2], Eq. (2.18), p. 364); a result equivalent to (3.9) was obtained by Saini [8].

4. The Relation for Current in Terms of Cauchy Data

In a previous paper [2], we showed that *in the noninductive case, the necessary and sufficient condition for coupling of electromagnetic and mechanical phenomena is that the current four-vector,* J^k, *be unspecified in the Cauchy problem.* Then, it follows that the unknown part of J^k lies along some vector, $'T^k$, tangent to the initial hypersurface S_3, which is of class C^2. In the corresponding discontinuity theory, the jump vector $[J^k]$ lies along some vector, T^k, tangent to a characteristic hypersurface S_3 of class C^2. We shall show that *the unknown part of* J^k, *in the case of the Cauchy problem for a self-inductive fluid, lies along* $''T^k$, *a properly chosen tangent vector.* Also, the new equations for the Cauchy problem are determined. In the following paragraphs, those results which follow from the same arguments as that of our previous paper [2] will merely be listed. For details of these arguments, the reader should see our previous paper [2]. Note, in discontinuity theory, it follows that in the self-inductive case, $[J^k]$ lies along some T^k.

Let us briefly review our notation. Along S_3, we assume that ρ, S, u_j, H_{jk}, and their tangential derivatives are *known* but their normal derivatives as well as J_k and all of its first derivatives are *unknown*. We denote S_3 by $\phi(x^j)$ = constant. Further, we write

$$\phi_j \equiv \frac{\partial\phi}{\partial x^j}, \qquad \phi^j \equiv g^{jk}\phi_k, \qquad \Phi \equiv \phi^j\phi_j \tag{4.1}$$

$$L \equiv \phi_j u^j, \qquad R \equiv \phi_j E^j. \tag{4.2}$$

Again, if $\underset{a}{t}{}^j$ (a = 1, 2, 3) are three mutually orthogonal unit vectors tangent to S_3 at each point, we can write [2]

$$\nabla_j u_k = \phi_j U_k + \sum_a \underset{a}{t}{}_j \underset{a}{A}{}_k \tag{4.3}$$

$$\nabla_j H_{kp} = \phi_j G_{kp} + \sum_a \underset{a}{t}{}_j \underset{a}{C}{}_{kp} \tag{4.4}$$

$$\nabla_j \rho = \phi_j P + \sum_a \underset{a}{t_j} \underset{a}{D} \tag{4.5}$$

$$\nabla_j \eta = \phi_j N + \sum_a \underset{a}{t_j} \underset{a}{E} \tag{4.6}$$

$$\nabla_j S = \phi_j M + \sum_a \underset{a}{t_j} \underset{a}{F} \ . \tag{4.7}$$

Evidently, U_k, G_{kp}, P, N, M are proportional to the unknown normal derivatives of u_k, H_{kp}, ρ, η, S, respectively. Similarly, $\underset{a}{A_k}$, $\underset{a}{C_{kp}}$, $\underset{a}{D}$, $\underset{a}{E}$, $\underset{a}{F}$ are *the known tangential derivatives* of u_k, H_{kp}, ρ, η, S, respectively. If we let

$$'a^2 = \left.\frac{\partial p}{\partial \eta}\right)_S , \qquad b = \left.\frac{\partial p}{\partial S}\right)_\eta \tag{4.8}$$

then it is easily shown that [2]

$$LM = K \tag{4.9}$$

$$\frac{\eta c^2}{\rho} P = (c^2 - {}'a^2) N - (b + \rho T) M + I \tag{4.10}$$

$$u_j U^j = 0 \tag{4.11}$$

$$(c^2 - {}'a^2) LN + \eta c^2 \phi_j U^j = {}'R \tag{4.12}$$

where K, I, $'R$ are known. The equations (4.11), (4.12) *are two of the basic equations for* U^j, N *of the Cauchy problem;* Eqs. (4.9), (4.10) determine M, P when N is known.

By use of (2.9), we shall analyze the Maxwell equations (2.13), (2.14). First, we follow our previous paper [2] and write

$$G_{pq} = \phi_p\, 'T_q - \phi_q\, 'T_p + \sum_{a>b} \underset{ab}{u} \left(\underset{a}{t_p} \underset{b}{t_q} - \underset{a}{t_q} \underset{b}{t_p} \right) \tag{4.13}$$

where u_{ab} are known. Use of (4.13), (4.4) shows that (2.13) is a condition on $C_{a\,pq}$. By differentiation of (2.9), we obtain

$$\mu\nabla_p F_{jk} = \nabla_p H_{jk} + (1-\lambda\mu)\Big(u^s u_k \nabla_p H_{sj} + H_{sj} u^s \nabla_p u_k + H_{sj} u_k \nabla_p u^s - u^s u_j \nabla_p H_{sk} - H_{sk} u^s \nabla_p u_j - H_{sk} u_j \nabla_p u^s\Big) . \tag{4.14}$$

Substituting (4.3), (4.4) [with G_{jk} replaced by (4.13)] into (4.14), and then inserting the resultant expression for $\nabla_p F_{jk}$ into (2.14), we obtain

$$\mu J^k = \Phi\,'T^k + (1-\lambda\mu)\Big[-\Phi\,'T^s u_s u^k + RU^k + H_{sj}\,\phi^j U^s u^k - u^s L(\phi_s\,'T^k - \phi^k\,'T_s) - E^k \phi_j U^j - H^{sk} LU_s\Big] + 'K^k \tag{4.15}$$

where $'K^k$ is known. By comparing (4.15) with the corresponding expression in the noninductive case (cf. [2], Eq. 3.20) — where $''K^k$ is known

$$\mu J^k = \Phi\,'T^k + ''K^k , \tag{4.16}$$

we see some of the difficulties of the self-inductive case.

Now, we list the *third and fourth of the basic relations for the Cauchy problem*. It should be noted that N, $'T^j$, U^j are considered as unknowns. First, we note that the *third basic relation* is

$$\phi_j\,'T^j = 0 . \tag{4.17}$$

Substituting (4.3), (4.6) into (2.11), we obtain by use of (4.2), (4.8)-(4.10)

$$\nabla_j M^{jk} = \eta LU^k + \eta\phi_j U^j u^k + (u^k L - 'a^2\phi^k)N + 'S^k \tag{4.18}$$

where $'S^k$ is known. Similarly, by substituting (4.3), (4.15) into (2.18), we obtain by use of (4.2) and the definitions [see (4.15)]

$$\delta = 1 - \lambda\mu \tag{4.19}$$

$$L^k = -\Phi\,'T^s u_s u^k + RU^k + H_{sj}\,\phi^j U^s u^k - u^s L(\phi_s\,'T^k - \phi^k\,'T_s) - E^k\phi_j U^j - H^{sk} L U_s\,, \tag{4.20}$$

the relation, where $'H^k$ is known,

$$\nabla_j E^{jk} = \mu H^{pk}(\Phi\,'T_p + \delta L_p) + \delta\mu H^{qp} u_q u^k(\Phi\,'T_p + \delta L_p) - \delta\left(F^{tq} H_{tq}\frac{g^{jp}}{4} - F^{jt} H_t^{\cdot p}\right) u^k \phi_j U_p - \delta\left(F^{tq} H_{tq}\frac{u^j}{4} + F^{qj} H_{pq} u^p\right)\phi_j U^k + {}'H^k. \tag{4.21}$$

By substituting (4.18), (4.21) into (2.10), we obtain the *fourth basic equation.* For infinite conductivity, the tensor E_{jk} admits a minimum of complications for the self-inductive case [see (2.9b)]. This case has been discussed by Saini [8].

To analyze (4.15), (4.21), we first note, by use of (4.1), (4.2), (4.17), that the *vector* L^k *of* (4.20) *is orthogonal to* ϕ^k. Thus, the equations (4.17), (4.15) can be written as

$$\phi_j\,''T^j = 0 \tag{4.21a}$$

$$\mu J^k = \Phi\,''T^k + {}'K^k \tag{4.21b}$$

where the tangent vector $''T^k$ to S_3 is defined by

$$\Phi\,''T^k = \Phi\,'T^k + \delta L^k\,. \tag{4.22}$$

Thus, again we see that *unknown component of* J^k *lies along a vector tangent to the initial hypersurface* S_3. By substituting (4.18), (4.21) into (2.10), we find by use of (4.22), (3.3)

$$U^k\left[\eta L + \delta\left(\frac{L}{4}F^{tq}H_{tq} + E_q F^{qj}\phi_j\right)\right] + u^k\left[\eta\phi_j U^j + \delta\left(\frac{F^{tq}}{4}H_{tq}\phi_j U^j - F^{jt}H_{tp}\phi_j U^p\right)\right] + N\left(Lu^k - {}'a^2\phi^k\right) + \Phi\,''T_p\left(\mu H^{pk} + \delta\mu H^{qp}u_q u^k\right) = {}'M^k \tag{4.23}$$

where $'M^k$ is known. Comparing (4.23) with the corresponding equation in the noninductive case (cf. [2], Eq. (3.23)), we see that F^{pk} *of the noninductive case has been replaced by* μH^{pk} *and three additional terms* (those multiplied by δ) *have been added.* The *four relations* (4.11), (4.12), (4.21a), (4.23) *are the basic system of seven linear algebraic equations (in nine unknowns,* U^j*,* $''T^j$*,* N*) for the Cauchy problem.*

REFERENCES

[1] R. Courant and K. O. Friedrichs, "Supersonic Flow and Shock Waves," p. 48. Interscience, New York, 1948.

[2] N. Coburn, Discontinuity relations for charged, compressible, relativistic fluids. *J. Math. and Mech.* 10, 361-392 (1961).

[3] P. M. Quan, Etude electromagnetique et thermodynamic d'unfluid relativiste charge. *J. Ratl. Mech. Anal.* 5, 473-538 (1956).

[4] A. Lichnerowicz, "Theories relativistes de la gravitation et de l'electromagnetisme." Masson, Paris, 1955.

[5a] J. L. Synge, "Relativity: The general theory." Interscience, New York, 1960.

[5b] J. L. Synge, "Relativity: The special theory." Interscience, New York, 1956.

[6] R. H. Abraham, Ph.D. thesis, University of Michigan, 1960.

[7] Y. Foures-Bruhat, Fluids charges de conductivite infinie. *Compt. rend. acad. Sci.* 248, 2558-2560 (1959).

[8] G. L. Saini, Singular hypersurfaces of order one in relativistic magneto-fluid-dynamics. *Proc. Roy. Soc.* A260, 61-78 (1961).

[9] R. Courant, "Methods of Mathematical Physics," Vol. II, Chapter V. Interscience, New York, 1962.

[10] A. H. Taub, Isentropic hydrodynamics for plane symmetric space-times. *Phys. Rev.* 103, 454-467 (1956).

SINGULAR PROPERTIES OF FLOW PATTERNS AT SMALL R_m*

G. S. S. Ludford

Cornell University
Ithaca, New York

This is a review of some magnetohydrodynamical problems which have been treated during the past four years.† As the title indicates, emphasis will be on the anomalies; and several questions will remain unanswered.

We are concerned with the flow of a slightly conducting fluid past a body, and for simplicity will consider only plane motions. The magnetic Reynolds number R_m is proportional to the conductivity and is therefore small. This means that the applied magnetic field is essentially unaffected by the flow, which is a major mathematical simplification.

The equations of steady motion for an incompressible, inviscid fluid are now, in nondimensional form,

$$\mathbf{v} \cdot \operatorname{grad} \mathbf{v} = -\operatorname{grad} p + N(\mathbf{E}_0 + \mathbf{v} \times \mathbf{H}_0) \times \mathbf{H}_0, \tag{1a}$$

$$\operatorname{div} \mathbf{v} = 0. \tag{1b}$$

The magnetic influence number

$$N = \frac{\mu^2 h^2 a\sigma}{\rho_0 U_0}$$

*This research has been supported partly by the National Science Foundation under Grant 19911 and partly by the Office of Naval Research under Contract Nonr-401 (46).

†Reference [5] gives an earlier review.

[μ = permeability, σ = conductivity, ρ_0 = density, U_0 = free-stream velocity, h = representative magnetic intensity, a = representative length] is a measure of the distortion of the flow pattern, from its potential form for $N = 0$, due to the magnetic field. The electric field $(\mu h U_0)\mathbf{E}_0$ is constant and perpendicular to the plane of $U_0\mathbf{v}$ and $h\mathbf{H}_0$.

We treat the extreme cases of N small and N large. The former holds for all moderate magnetic fields when σ is small; for the latter we require extremely strong fields.

N Small. If $\mathbf{v}_0$ is the potential flow past the body, we may expect that to a first approximation this flow is disturbed by the body force $\mathbf{F} = N(\mathbf{E}_0 + \mathbf{v}_0 \times \mathbf{H}_0) \times \mathbf{H}_0$, which is in general non-conservative.* The vorticity $\boldsymbol{\omega}$ thereby created may be obtained by integrating

$$\frac{d\boldsymbol{\omega}}{dt} \equiv v_0 \frac{\partial \boldsymbol{\omega}}{\partial s} = \operatorname{curl} \mathbf{F} \tag{2}$$

along a potential streamline s; and then the flow pattern itself follows from ω (at least in principle). Equation (2), which comes from the curl of (1a), has a single component, perpendicular to the plane of motion.

But it is easy to see that such a perturbation is not valid near the surface of the body. In potential flow, a particle takes a time $O(\log d)$ to pass by the front stagnation point, where d is its closest approach to the point, and during this time it acquires rotation steadily according to (2). Such particles later lie $O(d)$ from the surface, so that the vorticity becomes logarithmically infinite as the surface is approached.

The singularity arises at the front stagnation point, and we must look more carefully at the local solution of Eqs. (1) there. With $\mathbf{H}_0 = (\cos\phi, \sin\phi)$ and x measured along the surface, these equations possess the exact solution

$$u = -2a\cot\phi\, y + Cy^{1+k} + \alpha x, \qquad v = -\alpha y$$

where $k = N\sin^2\phi/\alpha$ and α, C are arbitrary constants. For $N = 0$ the first two terms in u combine to provide an arbitrary inclination for the dividing streamline, corresponding to

*This problem was treated by Ludford and Murray [3,4], for axially symmetric motion. Subsequently Murray and Chi [11] discussed the plane case.

the amount of vorticity in the incident stream. [In our case of a disturbed potential flow, $C = 2\alpha \cot \phi + O(N)$.] However, for $N \neq 0$ the ultimate angle of approach is ϕ irrespective of C. The pressure and inertia forces adjust to any angle, so that the magnetic field, however weak its effect, is able to force the fluid to move in its direction: $\mathbf{v} \times \mathbf{H}_0 = 0$ and the $\mathbf{E}_0 \times \mathbf{H}_0$ term is balanced by pressure in (1a).

The origin of our difficulty is now clear. The vorticity is

$$\omega = 2\alpha \cot \phi - C(1 + k)y^k$$

and in a simple perturbation theory we set

$$y^k = 1 + k \log y + O(N^2) ,$$

an expansion which is not valid near $y = 0$: ω is not infinite at the surface, but it is nonanalytic.

This raises the question of boundary layers for vanishingly small viscosity. In ordinary boundary-layer theory, applied, say, to an inviscid flow with uniform vorticity, the vorticity outside the boundary layer is changed by an exponentially small amount (in the Reynolds number). No disturbance of the vorticity in the main flow is caused by the viscous forces; the change is due to diffusion of the vortex sheet at the boundary. In the present case, however, the ponderomotive force creates a nonuniform vorticity in the inviscid flow which is redistributed by the viscous forces. The exponentially small diffusion of vorticity out of the boundary sheet is masked by this grosser algebraic effect.

An exact solution is available for stagnation-point flow [8], so that all this can be traced in detail. The analysis shows that difficult matching questions are bound to arise in any more general problem, especially on account of the curious power y^k, and a more extensive investigation is needed.

In this connection we note that the novel mathematical features are overlooked by those authors who treat similarity solutions. Such solutions, by their very structure, automatically exclude the y^k term and must be classed as exceptional.

N Large. The magnetic field must now be specified and we choose a uniform one, $h\mathbf{j}$, perpendicular to the undisturbed flow, $U_0\mathbf{i}$. For no current flow at infinity $E_0 = -\mathbf{i} \times \mathbf{j}$, so that in the limit $N \to \infty$ Eqs. (1) give

$$(\mathbf{v} - \mathbf{i}) \times \mathbf{j} = 0, \qquad \mathbf{v} = \mathbf{v}(x) .$$

The disturbance velocity is in the y-direction and depends only on x. For a body symmetric about the x-axis this determines the limit flow. Each y-column of fluid above and below the x-axis moves forward at constant velocity $\mathbf{i}$, slipping sideways to pass around the obstacle. The streamlines on each side are congruent and the disturbance extends to infinity.

In the nonsymmetric case there may be lateral slip before reaching the body or after leaving it, and we must go to the next approximation to determine this.* Here simple perturbation in N does not suffice.

At any fixed point changes in the y-direction vanish in the limit. To retain them we compress y as $N \to \infty$. The cumulative effect of the inertia terms is thereby assessed and undisturbed conditions at infinity regained. With disturbance velocity $\mathbf{v} - \mathbf{i} = (u,v)$ the required transformation is

$$Y = N^{-1/2}y, \qquad U = N^{1/2}u, \qquad P = N^{-1/2}p,$$

and elimination of all but v from the new limit equations yields

$$\frac{\partial^3 v}{\partial x^3} + \frac{\partial^2 v}{\partial Y^2} = 0. \tag{3}$$

If the body lies between the verticals $x = \pm 1$ the boundary conditions in the xY-plane are

$$v = f'_{\pm}(x) \qquad \text{on} \qquad Y = \pm 0 \qquad \text{for} \qquad |x| \leq 1.$$

Elsewhere v and $\partial v/\partial Y$ are continuous across the x-axis.† Here $y = f_{\pm}(x)$ is the equation of the upper/lower surface of the cylinder.

Equation (3) is parabolic and it is natural to consider the half plane $Y > 0$ with initial values prescribed on $Y = +0$. Then the problem is reduced to finding the source solution, that is, the x-derivative of the solution $\mathcal{H}$ which assumes unit-step

*Details are given in [6].

†This ensures the continuity of velocity and pressure.

initial values. $\mathcal{H}$ is a function of $\eta = x/Y^{2/3}$ only, whose precise form is not important for our present discussion. We need only note that it tends to one exponentially as $\eta \to +\infty$ but tends to zero algebraically as $\eta \to -\infty$. [For the heat equation $\eta = x/Y^{1/2}$, $\mathcal{H} = (1 + \operatorname{erf} \eta)/2$ and it is exponential in both cases.] As a consequence a source creates a disturbance directly upstream but not directly downstream of itself, which accounts for the one-sided integral equation mentioned below.

For a symmetric cylinder this is the problem, since $v = 0$ on $Y = +0$ for $|x| > 1$. Otherwise assume $v = V(x)$ on the remainder of the x-axis; write down the integral for v in terms of $\mathcal{H}$, f'_+ and V; and express the condition that it gives the same values of $\partial v/\partial Y$ there as are obtained from the corresponding integral in the lower half plane. The result is an Abel integral equation for $V(x)$, so that V is obtained explicitly.

This function determines the flow pattern in the whole xy-plane. For a flat plate at incidence α,

$$V = \frac{2 \tan \alpha}{\pi} \left[\sqrt{\frac{2}{-1-x}} - \tan^{-1} \sqrt{\frac{2}{-1-x}} \right] \quad \text{for} \quad x < -1, \tag{4}$$

while $V = 0$ for $x > 1$ as a consequence of purely upstream disturbance.

Since the complete flow region is doubly connected the question of uniqueness of the solution arises, just as it does in the absence of a magnetic field [Eqs. (1) with $N = 0$]. There seems little doubt of the uniqueness of the solution in either the upper or lower half plane, given $V(x)$ on the x-axis. We have, for a solution v in the upper half plane taking on zero initial values in a sufficiently regular manner,

$$0 = \int_{-\infty}^{\infty} dx \int_0^{\infty} v(v_{xxx} + v_{YY})\, dY = \int_0^{\infty} \left[v v_{xx} - \frac{1}{2} v_x^2 \right]_{x=-\infty}^{x=\infty} dY$$

$$+ \int_{-\infty}^{\infty} \left[v v_Y \right]_{Y=0}^{Y=\infty} dx - \int_{-\infty}^{\infty} dx \int_0^{\infty} \frac{1}{2} v_Y^2\, dY$$

$$= -\frac{1}{2} \int_{-\infty}^{\infty} dx \int_0^{\infty} v_Y^2\, dY$$

provided v vanishes sufficiently rapidly at ∞ [$O(1/r^{\epsilon})$ for arbitrary positive ϵ is sufficient]. Hence $v_Y = 0$ and $v = 0$. A precise uniqueness theorem would be desirable, however. The regularity condition must exclude solutions such as $\partial^3\mathfrak{H}/\partial x^3$, which tends to zero as $Y \to +0$ for each fixed x (cf. the heat doublet [2]).*

The real question of uniqueness arises at the integral equation for V which links the upper and lower solutions. But fortunately this is an Abel equation which is known to have a unique solution.

It would be of considerable interest to know more about Eq. (3), especially for an infinite region. Sichel [12, 13] has encountered this equation in another connection and has pointed out that its history extends back to Block [1] and Stadler [14]. It seems that there has been no specific treatment since then.

Yet another question is the next approximation. In general there are singularities on both $x = -1$ and $x = +1$ in the xy-plane,† and these worsen on perturbation. It may be that the singularities must be set on slightly swept-back curves, so as to take account of the wake of which the disturbance in $|x| \leq 1$ is the limiting form.

The corresponding three-dimensional problem has also been dealt with [7]. In this connection we note that a small R_m and large N may also be achieved by just taking U_0 small. The problem is then apparently equivalent to Stewartson's [15], which he treated by finding the approach to steady flow by means of linearized equations. But even when Stewartson's results are corrected [9, 10] they are different, and an explanation is given in [7].

REFERENCES

[1] H. Block, Sur les équations linéaires aux dérivées partielles à caractéristiques multiples. *Arkiv Mat. Astron. Fys.* 7, Nos. 13 and 21 (1912); 8, No. 23 (1912-13); 9, No. 8 (1913-14).

*Observations similar to those contained in this paragraph have been made by a referee and B. Friedman (private communications).

†The flat plate has only one (see [4]).

[2] H. S. Carslaw and J. S. Jaeger, "Conduction of Heat in Solids," p. 37. Oxford Univ. Press, London and New York, 1959.

[3] G. S. S. Ludford and J. D. Murray, Further results on the flow of a conducting fluid past a magnetized sphere. *Proc. 6th Midwestern Conf. Fluid Mech., Austin, Texas, 1959*, pp. 457-465 (1959).

[4] G. S. S. Ludford and J. D. Murray, On the flow of a conducting fluid past a magnetized sphere. *J. Fluid Mech.* 7, pp. 516-528 (1960).

[5] G. S. S. Ludford, Inviscid flow past a body at low magnetic Reynolds number. *Revs. Modern Phys.* 32, 1000-1003 (1960).

[6] G. S. S. Ludford, The effect of a very strong magnetic cross-field on steady motion through a slightly conducting fluid. *J. Fluid Mech.* 10, 141-155 (1961).

[7] G. S. S. Ludford, The effect of a very strong magnetic cross-field on steady motion through a slightly conducting fluid: three-dimensional case. *Arch. Ratl. Mech. Anal.* 8, 242-253 (1961).

[8] G. S. S. Ludford, Hydromagnetic stagnation-point flow for small R_m. *Z. angew. Math. Mech.* 43, 9-24 (1963).

[9] G. S. S. Ludford and M. P. Singh, The motion of a non-conducting sphere through a conducting fluid in a magnetic cross-field. *Proc. Cambridge Phil. Soc.* 59 (1963).

[10] G. S. S. Ludford and M. P. Singh, On the motion of a sphere through a conducting fluid in the presence of a magnetic field. *Proc. Cambridge Phil. Soc.* 59 (1963).

[11] J. D. Murray and L. K. Chi, The flow of a conducting fluid past a magnetized cylinder. *Mathematika* 7, 64-77 (1960).

[12] M. Sichel, A study of the leading edge of a shock-induced boundary layer. Ph.D. Thesis, Princeton University (1961). See also *Phys. Fluids* 5, 1168-80 (1962).

[13] M. Sichel, A study of the equation, $\phi_{xxx} - \phi_x \phi_{xx} + \phi_{yy} = 0$, which describes the structure of weak non-Hugoniot shock waves. Contract AF 49(638)-465, Princeton Univ. Rept. 541 (1961).

[14] G. Stadler, "Etudes sur l'Équation $\partial^p u/\partial x^p + a\Delta^m u = 0$." Lund, 1916.

[15] K. Stewartson, Motion of a sphere through a conducting fluid in the presence of a strong magnetic field. *Proc. Cambridge Phil. Soc.* 52, 301-316 (1956).

STABILITY OF MAGNETO-FLUID FREE BOUNDARIES AGAINST FINITE PERTURBATION*

A. A. Blank

New York University
New York, New York

1. Foreword

Instability has been a great bugbear to the latter day alchemists, those who seek to transmute sea water into a useful fuel. It would seem that if any way exists for a thermonuclear plasma to cooperate in its own destruction, the gas will invariably find it. In fact, the history of the battle against instability is a long and gloomy one, liberally strewn with the corpses of noble experiments. Out of these experimental failures there has developed a deeper theoretical understanding of the question of stability. Here we shall examine one of the most completely analyzed areas of the theory, the stability of an equilibrium characterized by a sharply defined free boundary between the conducting fluid and the electromagnetic field. Stable equilibria of this kind if attainable have great practical interest because they make it possible to attain fluid pressures as great as the magnetic pressure and because the dissipation

*The research presented in this paper is supported by the U. S. Atomic Energy Commission under contract AT(30-1)-1480.

of energy through "cyclotron" radiation is reduced to a surface effect.

The success of our stability investigations depends upon the use of a variational method rather than the traditional method of linear vibration analysis. With the variational approach we can handle finite amplitude perturbations as well as the infinitesimal vibration modes. Furthermore all the practically significant stable free boundaries have singularities in the form of cusps and normal mode analysis is inapplicable in the neighborhood of a cusp.

The mathematical studies of equilibrium and stability of magneto-fluids by variational methods were initiated by H. Grad in connection with the thermonuclear program of Project Sherwood. In 1954, Grad gave a variational description of fluid-magnetic equilibria for both the case of field and fluid separated sharply at an interface and the case in which the field and fluid interpenetrate [1]. At that time he gave the variational formulation for the stability of the free boundary. Shortly afterward, Grad obtained necessary and sufficient local conditions for stability. Arbitrary combinations of magnetic constraints in the form of constant currents and constant fluxes were treated. The stability criterion is largely independent of the constraints imposed, an extremely favorable fact for the practical utility of these results. Grad also formulated the local stability criterion in terms of a simple geometrical property of the interface [2]. This geometrical criterion eliminates the possibility that a finite fluid domain bounded by a smooth interface is stable. K. O. Friedrichs observed that the interface need not be a smooth surface, but can support cusped singularities where magnetic lines meet at a tangent. With this observation free boundary equilibria became practically interesting.

Grad's further investigations extended these results to a diverse class of fluid models ranging from incompressible fluids to a Boltzmann gas. An arbitrary additional conservative force field (such as gravity) was included in the analysis. It was shown that stability held against finite amplitude perturbations [3]. The research was essentially complete in 1956 but did not become generally available until 1958 when a cursory digest was published upon the declassification of Project Sherwood [4].

My own contribution was to complete some of the mathematical details, most notably with respect to the question of stability against finite amplitude perturbations. I shall attempt to go as directly as possible to that question, sketching in the necessary minimum of information as I go. A detailed report on the whole subject will be available shortly [5].

This variational approach has been extended beyond the class of problems treated here by Bernstein *et al.*, who treat the question of stability when the fluid and field interpenetrate [6]. The application of their results has so far proved to be difficult. The heart of the difficulty is that the question for stability for the intermixed case can not be reduced to a local property of the equilibrium configuration ([4], Section 10). In other words, widely separated parts of the fluid can cooperate in escaping from equilibrium. The literature of the interpenetrating case is large and rapidly growing. In particular, the special case of cylindrical symmetry has been extensively investigated by Rosenbluth, Suydam, and Taylor [7]. Rubin [8] has obtained a class of axially symmetric solutions of the nature of mirror machines.

2. Introduction

The free boundary equilibria which we consider are fluid magnetic configurations in which a perfectly conducting static fluid is sharply separated from a vacuum electromagnetic field by an interface (Fig. 1). The interface is permitted to carry surface currents and charges so that the electromagnetic field may have a finite jump at the interface. Our purpose

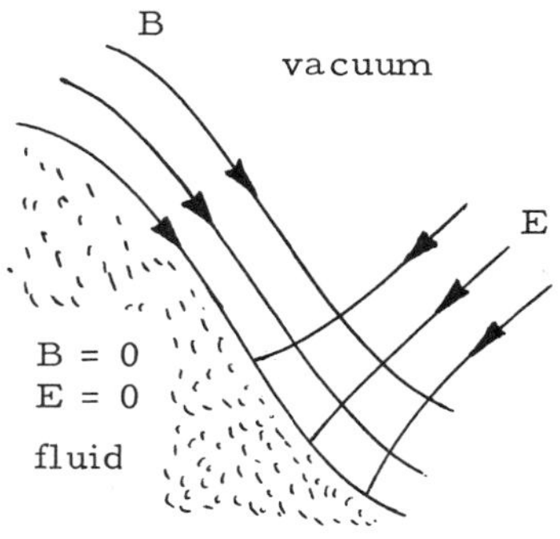

Figure 1

here is to obtain a criterion for stability and to find bounds for the amplitude of the perturbations against which the equilibrium is stable.

The sharp interface is an idealization and the practical utility of such equilibria depends on how closely that ideal is attainable (on intuitive grounds we may suppose that the boundary layer need be no thicker than an ion Larmor radius and may conceivably be thinner). Here we shall merely discuss the model and suppose that there is some valid range of application.

For a fluid system the concept of stability requires some explanation. If we were dealing with a Hamiltonian system with finitely many degrees of freedom we would have stability if the equilibrium corresponds to an absolute minimum of a convex potential. In that case a small perturbation can cause only a small deviation from equilibrium, and a finite perturbation, a correspondingly finite deviation. If the equilibrium corresponds to a relative minimum of the potential then there exists a bound to the energy perturbation below which the system will remain in a neighborhood of the equilibrium. In the fluid case the first of these conclusions remains valid, but the second may not. Consider, for example, the simple case of water held in a container by gravity (Fig. 2). There is an energy integral: the sum of the kinetic energy K and the gravitational energy P. We let P_0 denote the equilibrium potential. For the infinite vessel, Fig. 2a, P_0 is the absolute minimum of P. Given an energy perturbation ϵ, we have $K + P = P_0 + \epsilon$ and conclude without further reference to the equations of motion that

(a) (b)

Figure 2

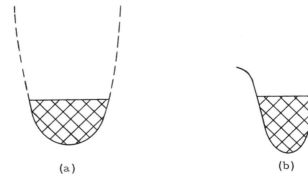

$$O \leq K \leq \epsilon, \qquad P_0 \leq P \leq P_0 + \epsilon. \tag{1}$$

The first inequality of (1) states that the kinetic energy is never large. The second inequality implies that the water remains close to the equilibrium in the sense that there are bounds on the amount of fluid which can exceed a given height. Intuitively, the same conclusion holds for Fig. 2b; namely, that only a certain amount of water can be lost over the rim, but it is impossible to deduce this result from energy considerations alone. In principle, it is possible for a drop of water to go over the rim and gain large kinetic energy which may be passed back to the water in the container. Similarly, in our problem, we shall show that a given equilibrium minimizes an appropriate potential among all neighboring states. In the absence of any known mechanism for transferring back the energy of an escaping fragment of the fluid we shall infer stability, but shall not be able to rigorously prove it. With this caveat in mind, if the potential of an equilibrium configuration is a minimum among all neighboring states, we shall say for brevity that it is stable.

The value of the variational approach lies in the elimination of detailed reference to the equations of motion and their solutions. We shall need only certain constants of the motion, energy and others. Unfortunately the method forces us to compare the equilibrium potential with the potentials of all neighboring configurations having the same constants of the motion, but we would like to compare it only with the potentials of configurations attainable by actual motions. The method is conservative in that it may fail to predict stability when it exists in fact.

Using an infinitesimal approach we shall exhibit a condition for a positive second variation. The existence of a positive second variation is generally insufficient to guarantee a relative minimum of the potential and, hence, is a less conservative criterion for stability. In the fluid magnetic problems considered here, however, we shall derive the condition for a minimum as a direct extension of the condition for a positive second variation.

3. Motion and Equilibrium of the Free Boundry

For a nonrelativistic fluid, in the absence of significant radiation pressure, the displacement current term may be dropped from Maxwell's equations [9]. The appropriate vacuum field equations then become

$$\left.\begin{aligned} \operatorname{curl} B &= 0, & \frac{\partial B}{\partial t} + \operatorname{curl} E &= 0 \\ \operatorname{div} B &= 0, & \operatorname{div} E &= 0. \end{aligned}\right\} \qquad (2)$$

At an interface moving with fluid velocity u we have the boundary conditions

$$B_n = 0, \qquad (E + u \times B)_t = 0 \qquad (3)$$

where the subscripts n, t denote normal and tangential components, respectively. The total domain consisting of vacuum and fluid is assumed to have a boundary consisting of fixed perfectly conducting surfaces satisfying the boundary conditions

$$B_n \text{ fixed} \quad (\partial B_n/\partial t = 0), \qquad E_t = 0. \qquad (4)$$

We shall, for simplicity, suppose that $B_n = 0$ on the fixed conductors also. To solve for the magnetic field in the electromagnetic system (2), it will be necessary only to prescribe B_n without explicit reference to the boundary conditions on E. The magnetic field is then uniquely determined at each instant by certain parameters called *periods* [10]. These periods may be fluxes defined by

$$\Phi_r = \int_{\Sigma_r} B \cdot dS \qquad (r = 1, \ldots, p), \qquad (5)$$

where Σ_r denotes a crosscut in the magnetic domain, that is, an open surface bounded by a curve lying on the boundary of the vacuum domain, or the periods may be external currents I_r defined by the mmf's

$$\mu I_r = \oint_{\Gamma_r} B \cdot dx \qquad (r = 1, \ldots, p), \qquad (6)$$

where Γ_r is a closed curve on the boundary of the vacuum domain which links Σ_r once. To prescribe external currents we have to alter the condition $E_t = 0$ in (4) by violating it across certain slits ([10], p. 121 ff; [5], Section 2.5). The integer p is the first Betti number of the domain. We may in fact, prescribe a mixture of currents and fluxes and we shall suppose that there exist p independent periods, the first of constant current type and the remainder of constant flux type:

$$\left.\begin{aligned} I_\alpha &= \text{constant} \qquad (\alpha = 1, \ldots, m) \\ \Phi_\beta &= \text{constant} \qquad (\beta = m+1, \ldots, p). \end{aligned}\right\} \tag{7}$$

Given B_n and the period constraints (7), B is determined uniquely by the position of the interface.

We take for granted that the motion of the fluid will be uniquely determined by its initial state and the forces acting upon it. For simplicity, we shall assume that the only force acting is the electromagnetic stress at the boundary.† For the electromagnetic system (2) the Maxwell stress consists of the magnetic stress alone (cf. [9]). In that case the Maxwell stress is normal to the interface and of magnitude $B^2/2\mu$. The magnetic stress must be balanced by a fluid stress of the same magnitude and we obtain the boundary condition

$$p_n = B^2/2\mu \tag{7a}$$

since the fluid stress reduces to its normal component p_n. Given the initial state of the fluid (position and velocity) it is intuitive that its subsequent motion is determined by the given boundary conditions and period constraints under the electromagnetic equations (2) and the equations of fluid motion. As yet, we do not have a mathematically complete demonstration of this assertion.

With the admission of constant current constraints the energy of the total system is not constant, but it is possible to find a constant of the motion comparable to the energy. For the rate of change of the magnetic energy we compute

---

†It is possible to include the effect of an additional conservative field. This is done in reference [5].

$$\dot{M} = \frac{d}{dt} \int_{V_m(t)} \frac{B^2}{2\mu} dV = \int_{V_m} \frac{\partial}{\partial t} \frac{B^2}{2\mu} dV + \oint_S \frac{B^2}{2\mu} u \cdot dS$$

where V_m is the time-varying magnetic domain, S is its boundary, and u the velocity of the boundary. From the energy relation for the electromagnetic equations (2) (cf. [9]),

$$\frac{\partial}{\partial t}\left(\frac{B^2}{2\mu}\right) + \operatorname{div}\left(\frac{E \times B}{\mu}\right) = 0$$

we obtain

$$\dot{M} = \oint_S \left\{\frac{1}{2\mu} B^2 u - \frac{E \times B}{\mu}\right\} \cdot dS.$$

From (3), we have

$$(E \times B) \cdot n = B^2 u \cdot n,$$

on the interface, S_m. Using the fact that $u = 0$ on the fixed conductors we have

$$\dot{M} = - \int_{S_m} \frac{1}{2\mu} B^2 u \cdot dS - \frac{1}{\mu} \int_{S_c} (E \times B) \cdot dS \tag{8}$$

where S_m is the free interface and S_c is the part of the boundary consisting of fixed conductors. With constant current constraints we may not suppose $E_t = 0$ everywhere on S_c. In that case, evaluation of the second integral in (8) yields†

$$\frac{1}{\mu} \int_{S_c} (E \times B) \cdot dS = - I_\alpha \dot{\Phi}_\alpha = - \frac{d}{dt} (I_\alpha \Phi_\alpha)$$

(cf. [5], Section 2.5) whence we obtain, finally,

†We use the summation convention for bilinear forms $I\Phi$ with the indices $\alpha = 1, \ldots, m$ over constant current constraints, $\beta = m + 1, \ldots, p$ over constant flux constraints.

$$\dot{M} = -\int_{S_m} \frac{1}{2\mu} B^2 u \cdot dS + \frac{d}{dt} I_\alpha \Phi_\alpha . \tag{9}$$

We replace the magnetic energy by the quantity

$$M^* = M - I_\alpha \Phi_\alpha \tag{10}$$

where, now,

$$\dot{M}^* = -\int_{S_m} \frac{1}{2\mu} B^2 u \cdot dS . \tag{11}$$

To obtain the energy balance for the whole system we now compute the rate of change of the fluid energy. Taking heat flow across the interface to be zero (and neglecting radiation), we obtain the energy equation ([9], (2.3))

$$\frac{d}{dt} \int_{V_f} \left(e + \frac{1}{2} \rho u^2\right) dV = -\int_{S_f} u^i p^{ij} \, dS_j \tag{12}$$

where V_f is the fluid domain and S_f its boundary. Here e is the internal energy per unit volume and p^{ij} is the fluid stress. At the boundary the fluid stress must be normal to balance the magnetic stress and we have by Eq. (7a)

$$\int_{S_f} u^i p^{ij} \, dS_j = \int_{S_f} p_n u \cdot dS = -\int_{S_m} \frac{1}{2\mu} B^2 u \cdot dS \tag{13}$$

$$= \frac{dM^*}{dt} .$$

From (12) and (13) it follows that

$$\frac{d}{dt} (K + \mathcal{E}^* + M^*) = 0$$

where

$$K = \int_{V_f} \frac{1}{2} \rho u^2 \, dV , \qquad \mathcal{E}^* = \int_{V_f} e \, dV . \tag{14}$$

We have found a constant of the motion comparable to the energy, namely,

$$K + \mathcal{E}^* + M^* = \text{constant}. \tag{15}$$

In summary we have as constants of the motion the energy integral (15), and the constant flux and current conditions (7). If the fluid is incompressible we take $\mathcal{E}^* = 0$ in (15) and include the total fluid volume among the constants of the motion.

We next turn our attention to the equilibrium state. For present purposes, this implies a fluid in static and thermal equilibrium†; that is,

$$p = \text{constant}, \qquad T = \text{constant} \tag{16}$$

where T denotes temperature. Letting p_0 and T_0 represent the equilibrium pressure and temperature, we define a fluid potential P_f:

$$P_f = \mathcal{E}^* + \int_{V_f} p_0\, dV = \int_{V_f} (e + p_0)\, dV. \tag{17}$$

It can be proved that for the equilibrium state $\dot{P}_f$ is stationary and $\ddot{P}_f$ is non-negative. Furthermore, it can be proved subject only to the constraint of mass conservation that the equilibrium potential P_f^0 is an absolute minimum ([5], Section 3.3). The proof relies upon the fact that entropy is nondecreasing. The only perturbations which leave P_f^0 unchanged are incompressible; that is, volume-preserving.

In equilibrium the pressure is scalar and the boundary condition (7a) reduces to

$$p_0 = B_0^2/2\mu ; \tag{18}$$

that is, the equilibrium condition p = constant implies that B^2 is constant on the interface. Since B is already uniquely determined by the conditions (3), (4), and (7) it is clear that not

†The condition of thermal equilibrium may be relaxed if only adiabatic perturbations are permitted ([5], Section 3.2).

every surface may serve as an equilibrium interface. The problem of determining the equilibrium interface happens to be formally identical with the classical free boundary problem for an incompressible irrotational fluid flow where the appropriate boundary condition is that the absolute velocity is constant on the unknown free surface.

We introduce a magnetic potential P_m:

$$P_m = M^* + \int_{V_m} p_0 \, dV = \int_{V_m} \left(\frac{1}{2\mu} B^2 + p_0 \right) dV - I_\alpha \Phi_\alpha \, . \tag{19}$$

We may then replace the energy integral (15) by

$$K + P_f + P_m = \text{constant} \, ; \tag{20}$$

here we have increased the integral by

$$\int_{V_f + V_m} p_0 \, dV \tag{21}$$

which is constant since the total domain $V_f + V_m$ is assumed to be fixed. As the total potential of the system we take

$$P = P_f + P_m \, . \tag{22}$$

Since the equilibrium fluid potential P_f^0 must be a minimum among admissible configurations, it will be sufficient for a proof of stability to prove that the equilibrium magnetic potential P_m^0 is a minimum subject to the stated constraints. In other words, we have completely separated the fluid and magnetic problems and we have reduced the question of stability to the study of the magnetic potential alone.

4. Variation of the Magnetic Potential

We have, for the magnetic potential P_m of (19)

$$\dot{P}_m = \dot{M}^* + \frac{d}{dt} \int_{V_m} p_0 \, dV \, .$$

Using the fact that u may differ from zero only on the interface S_m, we have

$$\frac{d}{dt}\int_{V_m} p_0\, dV = \oint_S p_0 u \cdot dS = \int_{S_m} p_0 u \cdot dS .$$

From (11) it follows that

$$\dot{P}_m = \int_{S_m} \left(p_0 - \frac{B^2}{2\mu}\right) u \cdot dS . \qquad (23)$$

In equilibrium, we have $p_0 = B^2/2\mu$ on the interface by (18). The requirement that P_m be stationary ($\dot{P}_m = 0$) is, then, a variational formulation of the condition of equilibrium under the stated constants. From (23) we compute (cf. [5], Section 4.1)

$$\ddot{P}_m = \int_{V_m} \frac{1}{\mu}\dot{B}^2\, dV - \int_{S_m} u_n^2 \frac{\partial}{\partial n}\left(\frac{B^2}{2\mu}\right) dS . \qquad (24)$$

The first integral in (24) is always positive, hence stabilizing. We already know that $\ddot{P}_f$ is non-negative. If the second integral in (24) is negative for all velocity perturbations u_n of the interface then the equilibrium is stable in the sense that the second variation $\ddot{P}$ of the total potential is positive. This will certainly be the case if B^2 always increases going from the interface into the magnetic domain [the normal in (24) points into the fluid]. If it should happen anywhere on the interface that $\partial/\partial n\,(B^2/2\mu) > 0$ we cannot directly infer instability because the other terms may compensate. Nonetheless, it is always possible to construct a perturbation in the neighborhood of a given point of the interface such that the sign of $\ddot{P}$ is the sign of $-\partial/\partial n\,(B^2/2\mu)$ at the point. This is the result from which all else follows and we state it formally:

Theorem 1. Under the criterion that the second variation of the potential be positive for all perturbations compatible with the constraints, a sufficient condition for a free boundary equilibrium to be stable is that $\partial/\partial n\,(B^2/2\mu) < 0$ at all points of the interface S_m. A necessary condition is that

$\partial/\partial n\ (B^2/2\mu) \leq 0$; that is, a sufficient condition for instability is that $\partial/\partial n\ (B^2/2\mu) > 0$ anywhere on S_m.

The special perturbation required for the proof of Theorem 1 can easily be described intuitively.† Ideally, it would be a variation which leaves B unchanged at every point so that we would have $\int \dot{B}^2\, dV = 0$. By choosing a perturbation of the interface which follows a flux tube and leaves the volume unchanged (Fig. 3) we could guarantee this and also fit the perturbation with an incompressible variation of the fluid. The

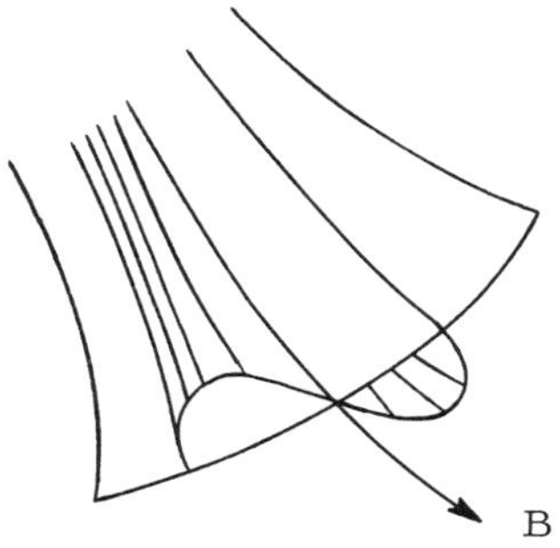

Figure 3

variation cannot follow a flux tube exactly for then it might lose its local character and extend over areas where both signs of $\partial/\partial n\ (B^2/2\mu)$ occur. The perturbation is modified so that it gradually tapers to zero along the length of the line. In such a construction we cannot make the first term of $\ddot{P}_m$ zero in (24), but we can make it arbitrarily small compared to the second term.

The criterion for stability in Theorem 1 has a simple geometrical interpretation. From the fact that $\nabla(B^2/2) = (B \cdot \nabla)B$ in the vacuum magnetic field it can be demonstrated that

$$\frac{\partial}{\partial n}\left(\frac{1}{2}\, B^2\right) = n \cdot (B \cdot \nabla)B = kB^2 \qquad (25)$$

†A detailed account is given in reference [5], Section A1.

where k is the curvature of the magnetic line. From the fact that B^2 is constant on the interface, Eq. (18), it follows that the principal normal of the magnetic line is normal to the surface (i.e., that the magnetic line is a geodesic). The sign of k is positive if the center of the curvature is on the fluid side of the normal and negative if it is on the vacuum side. We may therefore restate Theorem 1 as follows:

Theorem 2. Under the criterion that the second variation be positive, a sufficient condition for stability is that the centers of curvature of the magnetic lines on the interface always lie on the side of the vacuum; a reverse curvature anywhere is sufficient for instability.

As an immediate consequence of Theorem 2 we have:

Theorem 3. There exists no bounded stable fluid configuration separated from a magnetic field by an everywhere smooth interface.

This is intuitively evident because the magnetic line passing through the point of tangency of a plane of support would have to curve toward the fluid.†

For a free boundary configuration to be of interest in thermonuclear applications it follows that the interface must be singular in some respect. On the other hand, the fact that the bounded harmonic vector B cannot be singular, strongly restricts the nature of the singularities of the interface: these must be cusps where the magnetic lines meet at a tangent. The simplest of stable configurations of practical interest is an axially symmetric equilibrium such as might be maintained by a pair of opposed Helmholtz coils (Fig. 4).

5. Stability against Finite Perturbations

We should like to go beyond the infinitesimal variational analysis and demonstrate stability against finite amplitude perturbations of finite regions and even against disturbances of the whole boundary. This is desirable not only for the purpose of establishing bounds for the perturbations against which the equilibrium is stable, but also because the infinitesimal

---

†See reference [5], Section A2, for a detailed proof.

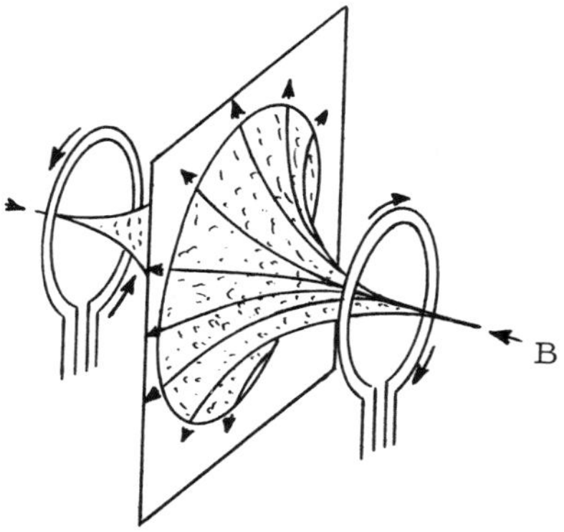

Fig. 4. [After Fig. 14.1(b) in A. Bishop, "Project Sherwood." Addison-Wesley, Reading, Massachusetts, 1958.]

variational analysis does not give us a way of handling the singularities of the interface. Since $B^2/2\mu = p_0$ on the interface and Theorem 1 requires that B^2 increase toward the vacuum for stability, we may naturally suppose that the magnetic potential P_m (and hence the total potential $P = P_m + P_f$) is a minimum if the quantity $B^2/2\mu - p_0$ is positive in a vacuum domain which contains the entire interface in its boundary. We shall, in fact, be able to prove that this is the case.

We let V_f^0 and V_m^0 denote the equilibrium fluid and magnetic domains and V_f and V_m the corresponding perturbed domains (the total domain $V = V_f^0 + V_m^0 = V_f + V_m$ is bounded by fixed conductors and remains constant). We suppose that there exists a subdomain $\overline{V}$ of V_m^0 which contains the entire equilibrium interface S_0 in its boundary and that

$$\frac{B_0^2}{2\mu} - p_0 > 0 \quad (\text{in } \overline{V}) \tag{26}$$

where B_0 denotes the equilibrium magnetic field. The perturbed field B is assumed to satisfy the same constraints (3), (4), and (7) as the equilibrium field. To compare P_m^0 and P_m we must find a method of extending the comparison from the volume which is covered by both domains to the volumes which are covered singly. We let V^+ denote that part of V_m which is not covered by V_m^0 and V^- denote that part of V_f which is not covered by V_f^0 (Fig. 5); that is,

$$V^+ = V_m \cap V_f^0 , \qquad V^- = V_f \cap V_m^0 .$$

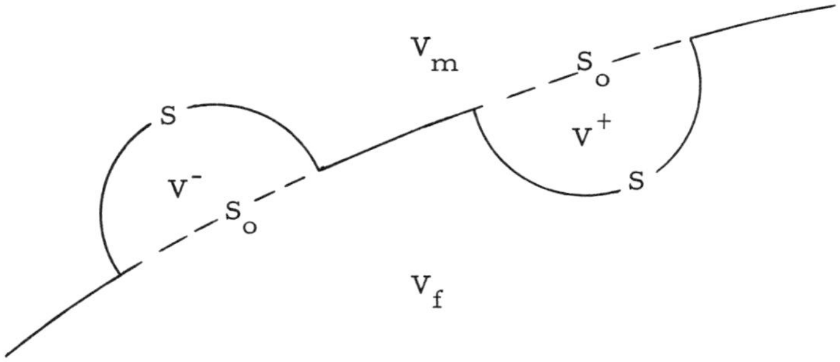

Figure 5

We suppose that V^- is contained in $\overline{V}$. To compare B_0^2 with B^2 in V^+ we shall construct a field X_0 which is equal to B_0 in V_m^0 and extends B_0 into V^+. Only the constant current constraints appear in the magnetic potential (19) and these must be satisfied by the extended field X_0. Letting ψ denote the potential of B in V_m^0 we extend ψ continuously and piecewise smoothly into V^+ and define

$$X_0 = \operatorname{grad} \psi \qquad (\text{in } V^+) . \tag{27}$$

In addition we require that

$$\frac{X_0^2}{2\mu} - p_0 \leq 0 \qquad (\text{in } V^+) . \tag{28}$$

Since $X_0 = B_0$ in V^- it follows from (26) and (28) that

$$\int_{V_m^0} \left\{ B_0^2/2\mu - p_0 \right\} dV \geq \int_{V_m} \left\{ X_0^2/2\mu - p_0 \right\} dV . \tag{29}$$

If we can express the harmonic field B as the minimizing vector in a class of admissible vectors X for a variational problem in V_m where the admissibility conditions are

$$\operatorname{curl} X = 0; \qquad I_\alpha[X] = I_\alpha \qquad (\alpha = 1, \ldots, m) \tag{30}$$

and where I_α denotes the given constant currents and by $I_r[X]$ we mean

$$I_r[X] = \frac{1}{\mu} \int_{\Gamma_r} X \cdot dx \qquad (r = 1, \ldots, p) , \tag{30a}$$

then we shall have effected the necessary comparison.

Elsewhere, we have shown how to generalize the Dirichlet principle to problems of this type ([10], Theorem 6). In this case, the functional to be minimized takes the form

$$F[X] = \int_{V_m} \frac{X^2}{2\mu} \, dV - \Phi_\beta I_\beta[X] \tag{31}$$

where Φ_β denotes the given constant fluxes. For X_0 we have

$$I_r[X_0] = \overset{0}{I_r} \qquad (r = 1, \ldots, p)$$

where $\overset{0}{I_r}$ designates the equilibrium values of the currents. The inequality $F[X_0] \geq F[B]$ takes the form

$$\int_{V_m} \frac{X_0^2}{2\mu} \, dV - \overset{0}{I_\beta} \Phi_\beta \geq \int_{V_m} \frac{B^2}{2\mu} \, dV - I_\beta \Phi_\beta . \tag{32}$$

In order to compare the potentials, the form of P_m given in (19) should be given in terms of a summation over β. By using the homogeneous boundary condition $B_n = 0$ in (4) on the fixed conductors we can do this quite simply. In that case, the magnetic energy can be put in the form ([10], Eq. (6.11))

$$M = \int_{V_m} \frac{B^2}{2\mu} \, dV = \frac{1}{2} I_r \Phi_r \tag{33}$$

whence

$$I_\alpha \Phi_\alpha = 2 \int_{V_m} \frac{B^2}{2\mu} \, dV - I_\beta \Phi_\beta .$$

Using this to rewrite the potential we obtain

$$P_m = I_\beta \Phi_\beta - \int_{V_m} \frac{B^2}{2\mu}\, dV + \int_{V_m} p_0\, dV \,. \tag{34}$$

From (29) it follows at once that $P_m \geq P_m^0$ and we have established that the equilibrium potential is a minimum.

A subdomain V^* of V_f^0 into which it is possible to extend B_0 under the conditions (27) and (28) will be called accessible. We summarize our results as follows:

Theorem 4. A sufficient condition that the equilibrium potential $P^0 = P_m^0 + P_f^0$ be less than that of a perturbed configuration is that the fluid boundary S remain within $\overline{V} + V^*$ where $\overline{V}$ denotes the part of V_m^0 contiguous to S_0 where $B_0^2/2\mu - p_0 > 0$ and V^* denotes the accessible part of $V^0{}_f$.

We have been able to show that every point of the interface lies within an accessible part of V_f^0. If the point is a regular point of the interface we may simply extend ψ inward by letting it be constant on a surface normal. In the planar case this construction even works in the neighborhood of a cusp. In that case we observe that ψ is proportional to arc-length measured from the cusp. We extend the normals inward from points of equal potential ψ_0 until they intersect and define $\psi = \psi_0$ on the broken line obtained in this way (Fig. 6). Clearly the same construction can be used for an axially symmetric configuration. In fact, for the important configuration of Fig. 4 which is axially symmetric and also has a plane of symmetry the construction shows that all of V_f^0 is accessible. We see no reason why this result should not be true in general.

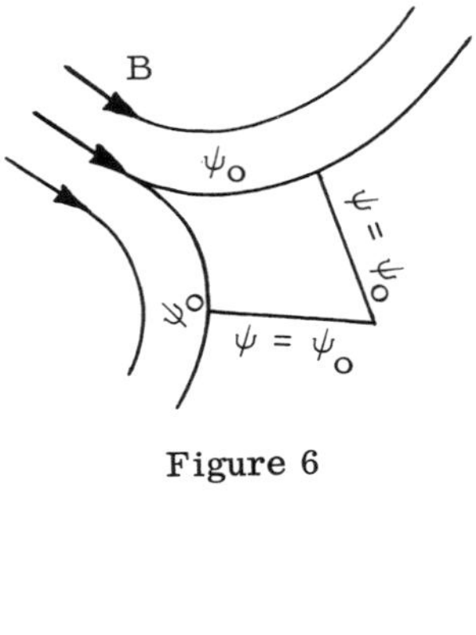

Figure 6

In the neighborhood of a cusped singularity in three dimensions the method of extension must be treated in a more sophisticated manner. In that case we extend ψ into the fluid by keeping it constant on a space involute to a B-line (Fig. 7). In other words, in the neighborhood of the cusp we find for each point x

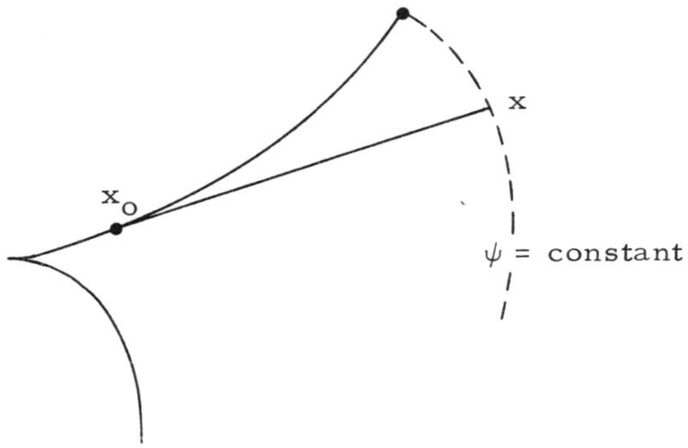

Figure 7

in V_f^0 the point x_0 on S_0 such that the tangent to the B-line at x_0 going away from the cusp passes through x.† Let the distance from x_0 to x be denoted by s. We assign at x the same value of ψ as at the point of the B-line through x_0 which lies at the distance s from x_0 in the direction away from the cusp.

To prove that these constructions actually fulfill our requirements is straightforward but tedious. The details are given in reference [5], Section A3. The two constructions described above may be patched together. In the method we use, the patches must meet on S_0 along an equipotential, ψ = constant.

A perturbation confined in V_f^0 to a domain V^* into which we may extend B_0 by (27) and (28) is called a regional disturbance. If V^* includes all of S_0 we say the disturbance is global. By means of the constructions we have described we can prove:

Theorem 5. If the condition $B_0^2/2\mu - p_0 > 0$ is satisfied in V_m^0 in the vicinity of the entire equilibrium interface S_0, then

†It is not difficult to show that the point x_0 exists and is unique.

the equilibrium is everywhere stable against regional disturbances. If $B_0^2/2\mu - p_0 < 0$ in any domain contiguous with the interface, then the equilibrium is unstable.

The second part of this theorem is proved in the same way as the corresponding part of Theorem 1.

We also have:

Theorem 6. If the condition $B_0^2/2\mu - p_0 > 0$ is satisfied everywhere in the vicinity of the interface and the geometry is plane or axially symmetric or such that the cusp lines are equipotentials for B_0 on the interface, then the equilibrium is stable against global disturbances.

REFERENCES

[1] AEC Rept. WASH-184, p. 144, Princeton (October 1954).
[2] AEC Rept. WASH-289, p. 115, Berkeley (February 1955).
[3] AEC Repts. TID-7503, p. 238, Princeton (October 1955); TID-7520, p. 373, Gatlinburg (June 1956).
[4] J. Berkowitz, H. Grad, and H. Rubin, *Proc. 2nd Intern. Conf. Peaceful Uses Atomic Energy, Geneva, 1958* 31, 171 (1959).
[5] A. A. Blank and H. Grad, Notes on Magneto-Hydronamics, IX. The motion and stability of fluid-magnetic free surfaces. NYO-6486, Courant Institute of Mathematical Sciences. New York University (1963).
[6] I. B. Bernstein, E. A. Frieman, M. D. Kruskal, and R. M Kulsrud, An energy principle for hydromagnetic stability problems. NYO 7315, Project Matterhorn, Princeton, New Jersey (1957); *Proc. Roy. Soc.* A244, 17-40 (1958).
[7] M. Rosenbluth, *Proc. 2nd Intern. Conf. Peaceful Uses Atomic Energy, Geneva, 1958* 31, 85 (1959); B. Suydam, *ibid.* p. 157; R. J. Taylor, *ibid.* p. 160. See also other papers by the same authors.
[8] H. Rubin, AEC Repts. TID-7503, p. 247, Princeton (1955); TID-7536 (Part 2), p. 193, Berkeley (1957).
[9] A. A. Blank and H. Grad, Notes on Magneto-Hydrodynamics VI. Fluid magnetic equations — general properties. NYO-6486, Magneto Fluid Dynamics Division, Courant Institute of Mathematical Sciences, New York University (1958).

[10] A. A. Blank, H. Grad, and K. O. Friedrichs, Notes on Magneto-Hydrodynamics, V. Theory of Maxwell's equations without displacement current. NYO-6486, Courant Institute of Mathematical Sciences, New York University (1957).

GENERALIZATION OF TOPOLOGICAL ASPECTS IN EXISTENCE PROOFS TO MAGNETOHYDRODYNAMICS

M. Z. v. Krzywoblocki

Michigan State University
East Lansing, Michigan

1. Introduction

In many applications of the electromagnetic equations, e.g., to fluid magnetics, the displacement current term in Maxwell's equations is considered negligibly small and therefore is omitted. The truncated electromagnetic system obtained in this way is called the pre-Maxwell system; it represents the contents of electromagnetic theory before the introduction of displacement current by Maxwell. It is possible to give a rather complete existence theory (i.e., existence of the solution) for the system of pre-Maxwell equations using some topological considerations. In fact, Blank *et al.* [1] accomplished this in a very elegant way. Their technique refers to two systems of two simultaneous equations: $\text{div}\ \mathbf{Z} = \sigma$, $\text{curl}\ \mathbf{Z} = \mathbf{S}$, where actually the scalar and vector magnitudes, σ and $\mathbf{S}$, respectively, are assumed to be given as functions of space and time. The vector function $\mathbf{Z}$ is the unknown variable. There are two such unknown variables. In the present work the author applies and generalizes these fundamentals and techniques to a system of equations of electromagnetohydrodynamics formulated in the macroscopic sense. The system in question consists actually of two main subsystems: the electromagnetic pre-Maxwell system (similar to the one mentioned above) and the hydrodynamic system referring to the motion of an inviscid, non-heat conducting and electrically conducting fluid. Due to the nature of the electromagnetohydrodynamic system of equations, there appear some additional equations auxiliary to the two main subsystems mentioned above. In the electromagnetic subsystem of equations these are: Ohm's law, Joule's heat, and the force arising from the charge density and from the induced magnetic effect due to the motion of the electrically conducting fluid through the magnetic lines of force. In the hydrodynamic subsystem of equations the additional equations consist of: equation of the conservation of energy, pressure-density-entropy relation and equation of state. Both main subsystems are of the form: $\mathrm{div}\ \mathbf{Z} = \sigma$, $\mathrm{curl}\ \mathbf{Z} = \mathbf{S}$; the hydrodynamic subsystem consists of two such sets: one referring to $\mathbf{A} = \rho\mathbf{q}$, another to $\mathbf{F} = \mathbf{q} \times \boldsymbol{\omega}$ (from Crocco's equation). In general, the magnitudes σ and $\mathbf{S}$ contain the unknown functions, i.e., $\mathbf{A}$ and $\mathbf{F}$. This is due to the nonlinearity of the equations of hydrodynamics. Since the magnitude of the current density depends (through the Ohm's law) upon the magnitude of the velocity vector $\mathbf{q}$, the magnitude $\mathbf{S}$ in one of the electromagnetic subset of equations is also a function of the unknown variables. To treat this system in the way analogous to that constructed by Blank, Friedrichs, and Grad, the following technique is proposed: the entire procedure is based upon a successive approximation process. In each step of the process in question the terms on the right-hand sides of the equations in the systems, given above ($\mathrm{div}\ \mathbf{Z} = \sigma$, $\mathrm{curl}\ \mathbf{Z} = \mathbf{S}$), supposedly contain the magnitudes of the unknown variables from the previous step. This enables one to apply all the constructed theorems without any difficulty. As a new item it is necessary to prove that the entire process, so constructed, converges. This is demonstrated in the following way: The solution of the system div-curl is proposed (by Blank, Friedrichs, and Grad) by means of the variational calculus. The author constructed, for illustrative purposes, the Euler-Lagrange differential equation. Under some assumption this is transformed into an integral equation and a general discussion on the subject of the convergence of the successive approximation procedure closes the work.

The fundamental idea in the present generalization is very simple. The system of equations governing the irrotational flow of an inviscid, non-heat-conducting incompressible fluid ($\mathrm{div}\ \mathbf{q} = 0$, $\mathrm{curl}\ \mathbf{q} = 0$), is a linear homogeneous system, identical to the pre-Maxwell system in a particular case, i.e., when the right-hand sides of the equations in the pre-Maxwell system are equal to zero. In the case of a rotational flow of an inviscid, non-heat-conducting but electrically conducting and magnetically permeable compressible fluid, the system of equations (div $\mathbf{q} = \mathbf{F}$, curl $\mathbf{q} = \mathbf{G}$) is nonlinear and nonhomogeneous. To deal with a linear nonhomogeneous system which automatically imbeds the entire system into the theory of Blank-Friedrichs-Grad, the author proposes the use of the technique of the successive approximations, with the right-hand sides of equations being the functions known from the previous steps of the procedure of the successive approximations. Such a procedure is usually a convergent one.

2. Fundamental Equations of Magnetohydrodynamics

Assume a three-dimensional nonsteady rotational flow of a perfect, inviscid, non-heat-conducting fluid in an electromagnetic field. The fundamental equations governing the hydrodynamic phenomena are the following:

equation of motion:

$$\mathbf{q}_{,t} + (\mathbf{q} \cdot \nabla)\,\mathbf{q} + \rho^{-1}\nabla p = \mathbf{P}\,; \tag{2.1}$$

equations of continuity and state:

$$\rho_{,t} + \nabla(\rho\mathbf{q}) = 0; \qquad p = R_1 \rho T\,; \tag{2.2}$$

pressure-density-entropy relation:

$$p_0^{-1} p = \left(\rho_0^{-1}\rho\right)^{\gamma} \exp\left[c_v^{-1}(S - S_0)\right], \tag{2.3}$$

where $\mathbf{P}$ denotes the external (including electromagnetic) forces for unit mass, $\mathbf{q}$ the velocity, and S the entropy. From the first law of thermodynamics written in the form

$$dQ = T\,dS = c_v\,dT + p\,d(\rho^{-1})\,,$$

with $$h = c_p T = c_v T + p\rho^{-1}, \qquad c_v\,,\,c_p = \text{constant},$$

one obtains the vector equation

$$T\nabla S - \nabla h + \rho^{-1}\nabla p = 0 ; \qquad T\nabla S = \nabla Q . \tag{2.4}$$

The expression dQ contains the Joule heat as well as any other energy (heat) addition or subtraction from or to outside. By adding to both sides of Eq. (2.1) the expression $\mathbf{q} \times (\nabla \times \mathbf{q})$, combining terms on the left-hand side and using Eq. (2.4), one obtains the generalized Crocco equation

$$\mathbf{q}_{,t} + \nabla H - \nabla Q = \mathbf{q} \times \omega + \mathbf{P} ; \tag{2.5}$$

$$H = \frac{1}{2} q^2 + h ; \qquad \omega = \text{curl}\, \mathbf{q} ; \tag{2.6}$$

obviously $H_0 = h_0 = c_p T_0$ at the point where $q = 0$. The velocity of sound is defined as

$$a^2 = (\partial p/\partial \rho)_S , \tag{2.7}$$

which in a steady flow can be calculated from the generalized energy equation of the form

$$\frac{1}{2} q^2 + (\gamma - 1)^{-1} a^2 - \int_0^s \mathbf{P} \cdot d\boldsymbol{\sigma} - \int_0^s dQ = K(\psi, \mu) , \tag{2.8}$$

where the integration is performed along a streamline, $(\psi, \mu) =$ constant, $\psi, \mu =$ streamsurfaces, from a point $s = 0$ to a point "s." The streamline represents an intersection of two streamsurfaces. Obviously, the function $K(\psi, \mu)$ assumes a constant value along a streamline, $\psi, \mu =$ constant. In the case of an unsteady flow one may calculate the velocity of sound from the generalized equation of energy obtained by means of the combination of Eqs. (2.1) and (2.4) and integration with respect to the running coordinate w along the particle line $[(\lambda, \chi, \nu) =$ constant]:

$$\frac{1}{2} q^2 + \int_0^w \mathbf{q}_{,t} \cdot d\mathbf{w}_1 + (\gamma - 1)^{-1} a^2$$

$$- \int_0^w \mathbf{P} \cdot d\mathbf{w}_1 - \int_0^w dQ = L(\lambda, \chi, \nu) . \tag{2.9}$$

The function $L(\lambda,\chi,\nu)$ assumes a constant value along a particle line, (λ,χ,ν) = constant. The "particle surfaces" (functions), λ,χ,μ, satisfy identically the continuity equation (2.2). In the case of an one-dimensional unsteady flow, the continuity equation, $\rho_{,t} + (\rho u)_{,x} = 0$, is identically satisfied by the particle function $\chi(x,t)$: $\rho = \chi_{,x}$; $\rho u = -\chi_{,t}$. The function χ represents a surface in the (x,t,χ)-space. In the case of an unsteady flow the resultant system of equations is of the form

$$\operatorname{div}(\rho\mathbf{q}) = -\rho_{,t}; \tag{2.10}$$

$$\operatorname{curl}\mathbf{q} = \boldsymbol{\omega}; \tag{2.11}$$

$$\mathbf{F} \equiv \mathbf{q}\times\boldsymbol{\omega} = \mathbf{q}_{,t} + \nabla(H-Q) - \mathbf{P}; \tag{2.12}$$

$$\frac{1}{2}\mathbf{q}^2 + \int_0^W \mathbf{q}_{,t}\cdot d\mathbf{w}_1 + (\gamma-1)^{-1}a^2$$

$$+ \int_0^W \mathbf{P}\cdot dw_1 - \int_0^W dQ = L(\lambda,\chi,\nu). \tag{2.13}$$

The function $L = L(\lambda,\chi,\nu)$ is a known and given function of the position and time. In the case the functions Q and $\mathbf{P}$ refer also to the electromagnetic phenomena, the corresponding electromagnetic quantities are known from the set of electromagnetic equation of the dynamic system in question. In this case the system of four equations (2.10) to (2.13) defines four functions

$$\mathbf{q},\ \rho,\ \boldsymbol{\omega}\ \text{ and }\ a^2 = \gamma R_1 T = \gamma p\rho^{-1}\ ,\ \text{ or }\ p\ \text{ or }\ T.$$

Some remarks, concerning the functions Q and $\mathbf{P}$ are given below.

Consider the following two fundamental equations (continuity and Crocco):

$$\operatorname{div}(\rho\mathbf{q}) + \rho_{,t} = \rho\operatorname{div}\mathbf{q} + \operatorname{grad}\rho\cdot\mathbf{q} + \rho_{,t} = 0; \tag{2.14}$$

$$\mathbf{q}\times\boldsymbol{\omega} = \mathbf{q}_{,t} + \nabla(H-Q) - \mathbf{P}. \tag{2.15}$$

From the Crocco equation we get

$$\mathbf{q} \times (\mathbf{q} \times \omega) = (\mathbf{q} \cdot \omega)\,\mathbf{q} - q^2 \omega = \mathbf{q} \times [\mathbf{q}_{,t} + \nabla(H - Q) - \mathbf{P}] . \qquad (2.15a)$$

We may propose a certain number of hydrodynamical systems, more or less useful:

$$\text{(I)} \quad \begin{cases} \operatorname{div} \mathbf{q} = -\rho^{-1} (\operatorname{grad} \rho \cdot \mathbf{q} + \rho_{,t}); \\ \operatorname{curl} \mathbf{q} = q^{-2}\left\{(\mathbf{q} \cdot \omega)\,\mathbf{q} - \mathbf{q} \times [\mathbf{q}_{,t} + \nabla(H - Q) - \mathbf{P}]\right\} ; \end{cases}$$

$$\text{(II)} \quad \begin{cases} \operatorname{div} (\rho\mathbf{q}) = -\rho_{,t}; \\ \operatorname{curl} (\rho\mathbf{q}) = \rho q^{-2}\left\{(\mathbf{q} \cdot \omega)\,\mathbf{q} - \mathbf{q} \times \left[\mathbf{q}_{,t} + \nabla(H - Q) - \mathbf{P}\right]\right\} \\ \qquad\qquad + \operatorname{grad} \rho \times \mathbf{q} ; \end{cases}$$

$$\text{(III)} \quad \begin{cases} \operatorname{div} (\mathbf{q} \times \boldsymbol{\omega}) = \operatorname{div} \mathbf{q}_{,t} + \nabla^2(H - Q) - \operatorname{div} \mathbf{P} ; \\ \operatorname{curl} (\mathbf{q} \times \omega) = \operatorname{curl} \mathbf{q}_{,t} - \operatorname{curl} \mathbf{P} . \end{cases}$$

For an incompressible, inviscid fluid in a steady irrotational flow the system (I) furnishes: $\operatorname{div} \mathbf{q} = 0, \ \operatorname{curl} \mathbf{q} = 0$.

One can propose various other hydrodynamical systems, depending upon the operations applied to the fundamental equations. Obviously, the above given systems may be used either in some combinations or alone. For the purpose of the particular interest in the present work, i.e., an inviscid, non-heat conducting fluid, it is sufficient to take into account only the system (I). But in other kinds of fluids, like the well-known viscous and heat-conducting fluid (call it ordinary viscous fluid), or the Stokesian fluid, the Reiner-Rivlin fluid, the Maxwellian fluid, etc., one may be compelled to use a combination of more complicated systems, analogous to the system (II) or even to the system (III), obviously with the addition of the terms involving viscosity, etc. The system (I) is very elegant, and the question of obtaining analogous forms in the case of more complicated models of fluids (like those mentioned above) reduces actually to the problem of deriving an equation for the vorticity vector in an explicit form. The problem of this nature is not a simple one. Although a great number of works was written on the subject of the vorticity in viscous fluids (for illustrative purposes see some of the papers by Truesdell

[6], Carstoiu [4]) the explicit expressions for the vorticity vector in viscous fluids are very complicated. Much simpler results can be obtained for the curl of the acceleration.

In the case of an unsteady flow of an inviscid, non-heat conducting fluid the fundamental hydrodynamical system of equations may be represented in the form

$$\operatorname{div} \mathbf{q} = W; \qquad W = -\rho^{-1}(\operatorname{grad} \rho \cdot \mathbf{q} + \rho_{,t}); \tag{2.16}$$

$$\operatorname{curl} \mathbf{q} = \mathbf{Z}; \qquad \mathbf{Z} = q^{-2}\left\{(\mathbf{q} \cdot \omega)\,\mathbf{q} - \mathbf{q} \times [\mathbf{q}_{,t} + \nabla(H - Q) - \mathbf{P}]\right\}, \tag{2.17}$$

with the functions W and $\mathbf{Z}$ supposedly being known and given. After the velocity vector is determined, one can calculate the a^2 (or T or p or ρ) from Eq. (2.13).

The electromagnetic part of the set of equations governing the dynamic system in question consists of two groups. As the first group we choose the pre-Maxwell system

$$\operatorname{curl} \mathbf{B} = \mu \mathbf{J} \tag{2.18a}$$

$$\operatorname{div} \mathbf{B} = 0; \tag{2.18b}$$

$$\operatorname{curl} \mathbf{E} = -\partial \mathbf{B}/\partial t \tag{2.19a}$$

$$\operatorname{div} \mathbf{E} = \kappa^{-1}\alpha, \tag{2.19b}$$

where the used symbols denote: $\mathbf{E}$ = electrostatic vector field intensity; κ = permittivity of the medium (dielectric constant); $\mathbf{H}$ = magnetic vector field intensity (magnetic flux density); $\mathbf{B} = \mu\mathbf{H}$; μ = permeability of the medium (assumed to be equal to unity); α = charge density; I = current (amp); $I = JA$; J = conduction current density; A = cross-sectional area.

These are two systems of equations, Eqs. (2.18), (2.19), in two unknown vectors $\mathbf{B}$ and $\mathbf{E}$.

The second group of the electromagnetic equations consists of the following equations:

(a) Ohm's law:

$$\mathbf{J} = \sigma_1(\mathbf{E} + \mu \mathbf{q} \times \mathbf{H}); \tag{2.20}$$

σ_1—electrical conductivity;

(b) Joule's heat:

$$Q_1 = J^2 \sigma_1^{-1}; \qquad Q = Q_1 + Q_{ext}; \tag{2.21}$$

(c) the force arising from the charge density and from the induced magnetic effect due to the motion of the electrically conducting fluid through the magnetic lines of force:

$$\mathbf{P}_1 = \alpha \mathbf{E} + \mathbf{J} \times \mathbf{B}; \qquad \mathbf{P} = \mathbf{P}_1 + \mathbf{P}_{ext}. \tag{2.22}$$

As it is seen, both systems, the hydrodynamic one, Eqs. (2.14) to (2.17), and the electromagnetic one, Eq. (2.18) to (2.19), are interrelated [i.e., the right-hand sides of each system contain the magnitudes (functions) which represent the unknowns in another system, and vice versa]. Moreover, the right-hand sides of each system contain the unknown functions from itself. Thus the functions $\mathbf{W}$ and $\mathbf{Z}$, Eqs. (2.16) and (2.17), contain the unknown functions ρ, $\mathbf{q}$, $\mathbf{J}$, $\mathbf{E}$, $\mathbf{H}$. The vector $\mathbf{J}$ in Eq. (2.18a) contains the unknown vectors $\mathbf{E}$, $\mathbf{q}$ and $\mathbf{H}$. The functions H, Q_{ext}, α, the vector $\mathbf{P}$, and the constant parameters κ, μ, σ_1 are known and given. All the other scalar functions and vectors can be calculated from the equations given above. This suggests that one may apply some sort of successive approximation process, in which the right-hand sides would consist of functions known and given from the previous steps of the successive approximation procedure. This idea is applied below. In each step the study of the pre-Maxwell system or of the hydrodynamical system is reduced to the study of systems having the form:

$$\operatorname{div} \mathbf{Z} = \sigma; \qquad \operatorname{curl} \mathbf{Z} = \Xi; \qquad \operatorname{div} \Xi = 0. \tag{2.23}$$

By comparison with Eqs. (2.14) to (2.17), we see that both systems of equations, hydrodynamical one, Eqs. (2.14) and (2.17), and electromagnetic one, Eqs. (2.18), (2.19) are actually representable in each step of the successive approximation procedure by means of the system (2.23). This system will be investigated below.

One can solve the system (2.18), (2.19) by first solving for $\mathbf{E}$ in the system:

$$\operatorname{curl} \operatorname{curl} \mathbf{E} = -\mu \partial \mathbf{J}/\partial t; \qquad \operatorname{div} \mathbf{E} = \kappa^{-1} \alpha, \tag{2.24}$$

obtained by eliminating $\partial \mathbf{B}/\partial t$ from Eq. (2.18a); defining a well-posed problem for this system and determining uniquely $\mathbf{E}$ as its solution, enables one to calculate $\mathbf{B}$ from Eq. (2.19a).

A current is defined by the period $(1/\mu)\oint \mathbf{B}\cdot d\mathbf{x}$ defined on a closed curve; a flux by $\int \mathbf{B}\cdot d\mathbf{S}$ over a disk; the total charge within a surface by $\kappa\oint \mathbf{E}\cdot d\mathbf{S}$; electromotive force (emf) along an arc by $\int \mathbf{E}\cdot d\mathbf{x}$.

As it is seen, the fundamental philosophy in the present generalization is the following: the system of equations of an irrotational flow of an inviscid incompressible fluid ($\operatorname{div}\mathbf{q} = 0$, $\operatorname{curl}\mathbf{q} = 0$) is identical to the system of equations of electromagnetics in special conditions ($\operatorname{div}\mathbf{B} = 0$; $\operatorname{curl}\mathbf{B} = 0$; or $\operatorname{div}\mathbf{E} = 0$; $\operatorname{curl}\mathbf{E} = 0$). The generalization to the electromagnetohydrodynamics of a rotational flow of a compressible fluid and embedding it into the entirety of the Blank-Friedrichs-Grad theory is obtained by a successive approximation procedure, in each step of which the right-hand sides of equations are assumed to be known and given.

3. Previous Fundamental Results

The system of equations fundamental to systems of equations (2.16), (2.17) and (2.18), (2.19) is of the following form [Eq. (2.23)]:

$$\operatorname{div}\mathbf{Z} = \sigma; \qquad \operatorname{curl}\mathbf{Z} = \Xi; \qquad \operatorname{div}\Xi = 0. \tag{3.1}$$

The system was extensively studied by Blank, Friedrichs, and Grad in connection with application of it to the pre-Maxwell system. Below, we collect the most important results of those investigations ([1], p. 21 ff.). Many of the aspects in the solution of the inhomogeneous system (3.1) may be assigned to the solution of the homogeneous system*:

$$\operatorname{div} Z = 0; \qquad \operatorname{curl} Z = 0. \tag{3.2}$$

The classical integral theorems of Gauss and Stokes form the most natural way of approach to the formulation of well-posed† boundary value problems for the systems (3.1) and (3.2):

---

*Beginning now, all vectors will be denoted by italics rather than bold face, since the notation is self-explainable.

†If a stated problem admits precisely one solution it is said to be a well-posed one.

$$\int_C \operatorname{grad} \phi \cdot dx = \phi(x_2) - \phi(x_1); \tag{3.3}$$

$$\int_\Sigma \operatorname{curl} Z \cdot dS = \oint_\Gamma Z \cdot dx; \tag{3.4}$$

$$\int_D \operatorname{div} Z \, dV = \oint_S Z \cdot dS. \tag{3.5}$$

Here the used symbols denote: C an open arc, Γ a closed curve, Σ an open surface, D a three-dimensional domain, S a closed surface.

For an irrotational vector Z in D one has

$$\oint_\Gamma Z \cdot dx = 0$$

provided Γ lies entirely in D, is closed, and bounds a surface Σ in D. To each Γ which does not bound in D we may associate the class of all oriented closed curves Γ' such that $\Gamma - \Gamma'$ bounds a surface in D; this implies that

$$\int_{\Gamma'} Z \cdot dx = \int_\Gamma Z \cdot dx,$$

and the value of the line integral does not depend upon Γ but upon this class, the homology class of Γ. This constant of a homology class is called a period of the vector field.* In an analogous way the divergence theorem yields a concept of

---

*We use the following periods:

$$[\Gamma] = [\Gamma; Z] = \oint_\Gamma Z \cdot dx \, ; \qquad \Gamma = \text{closed curve};$$

$$[S] = [S; Z] = \oint_S Z \cdot dS \, ; \qquad S = \text{closed surface};$$

$$[\Sigma] = [\Sigma; Y] = \int_\Sigma Y \cdot dS \, ; \qquad \Sigma = \text{open surface};$$

$$[C] = [C; X] = \int_C X \cdot dx \, ; \qquad C = \text{open curve}.$$

homology for surfaces when applied to solenoidal vectors. Two surfaces S and S' belong to the same homology class if $S-S'$ bounds a subdomain in D.

Assume that the three-dimensional D is orientable and possesses a boundary consisting of finitely many regular bounded closed surfaces. The topology is finite, i.e., there are only finitely many independent homology classes of each dimension. A closed curve is homologous to zero (~ 0) if it bounds a surface Σ in D. Two curves Γ and Γ' are homologous if $\Gamma - \Gamma' \sim 0$. By the introduction of zero element the set of curves forms an additive group; this contains the smaller additive group of closed curves. The homology classes are additive, and the additive zero is the class of all curves homologous to zero. Integration over a linear combination of curves is defined as

$$\int_{\lambda C_1 + \mu C_2} Z \cdot dx = \lambda \int_{C_1} Z \cdot dx + \mu \int_{C_2} Z \cdot dx;$$

for $\operatorname{curl} Z = 0$, the expression

$$\int_{\Gamma} Z \cdot dx$$

is a linear function on the linear vector space of homology classes of closed curves, called homology group of closed curves (the basis of the space is finite).

A surface is homologous to zero if it bounds a subdomain in D. Assume that the boundary S^* of D consists of $q+1$ closed surfaces $S_0^*, S_1^*, \cdots, S_q^*$, oriented so that $S_0^* + S_1^* + \cdots + S_q^* \sim 0$, the sum bounding D. A basis for the homology group can be formed from any q of the components of the boundary.

We use the notation

$$[\Gamma] = [\Gamma; Z] = \oint_{\Gamma} Z \cdot dx; \qquad [S] = [S; Z] = \oint_{S} Z \cdot dS. \tag{3.6}$$

For a solenoidal vector Y, $\operatorname{div} Y = 0$, $[S;Y]$ vanishes for any $S \sim 0$. The function $[S;Y]$, defined on the class of closed surfaces, has the same value for each surface S in a given homology class. The constant associated in this way with a homology class defines the period. For an irrotational vector X, curl $X = 0$ in D, the function $[\Gamma;X]$ is defined on homology classes independently of the choice of individual curves; obviously, $\operatorname{div} Y = 0$ ($\operatorname{curl} X = 0$) if $[S;Y]$ ($[\Gamma;X]$) is constant on each homology class. Two surfaces (arcs) are homologous if

$$[S_1;Y] = [S_2;Y]([\Gamma_1;X] = [\Gamma_2;X])$$

for every solenoidal Y (irrotational X). Another requirement to define homology of two closed curves (surfaces) is that the two integrals $[\Gamma;X]$ ($[S;Y]$) are the same for each irrotational X (solenoidal Y). The requirement $\operatorname{curl} X = 0$ furnishes the existence of a single-valued scalar potential ($X = \operatorname{grad} \phi$, $\operatorname{curl} \operatorname{grad} \phi = 0$) locally. If we require the existence of a potential ϕ over all D, then it is both necessary and sufficient that the periods $[\Gamma;X]$ all vanish. For any solenoidal vector Y there exists locally a vector potential A, $Y = \operatorname{curl} A$, $\operatorname{div} \operatorname{curl} A = 0$. Again, it is both necessary and sufficient condition for A being a single valued in the whole D that the closed periods $[S;Y]$ all vanish. The topology is finite, which implies that each closed curve and surface is homologous to a linear combination of a given set of basis elements:

$$\Gamma \sim a_1\Gamma_1 + a_2\Gamma_2 + \cdots + a_p\Gamma_p; \qquad S \sim b_1S_1 + \cdots + b_qS_q , \tag{3.7}$$

where $\Gamma_i(S_i)$ denote a basis of the closed curve group (surface group). Obviously, when $\operatorname{curl} X = 0$, $\operatorname{div} Y = 0$, then all periods are defined in terms of the finite set of periods of the basis elements:

$$[\Gamma;X] = a_1[\Gamma_1;X] + \cdots + a_p[\Gamma_p;X] ; \tag{3.8}$$

$$[S;Y] = b_1[S_1;Y] + \cdots + b_q[S_q;Y] . \tag{3.9}$$

Above, it was demonstrated that for an irrotational vector the integral $\oint_\Gamma X \cdot dx$ depends only upon the homology class of Γ; for a solenoidal Y, the integral $\oint_S Y \cdot dS$ also depends only upon homology class. Below we consider the homogeneous differential equations supplemented with appropriate boundary conditions.

Assume an X in D with† $X_t = 0$ on the boundary S^*, and choose a basis for the homology group of closed curves in D on S^*; then $[\Gamma;X] = 0$ for all closed curves in D. Let us choose two arcs C and C' which begin and terminate on S^*; assume that one can find arcs on S^* which taken together with C and C' form a closed curve Γ; with $X_t = 0$ the only contribution to $[\Gamma;X]$ comes from C and C'; clearly, the integrals on C and C' are equal; the open arcs C' which with C are completed to form Γ by some additions from S^* constitute a class upon which the line integral $\int_{C'} X \cdot dx$ is constant (this second kind of homology is called homology modulo the boundary).

The Alexander duality theorem states that in a three-dimensional space the number of independent elements of the k-dimensional homology group of D is equal to that of the $(3-k-1)$-dimensional group of $\bar{D}$ ($\bar{D}$ complementary to D is the exterior domain to D). These two groups are called dual to each other. Thus the closed curve group in D is dual to the closed curve group in $\bar{D}$. The homology group of closed curves on the common boundary S^* is the direct sum of the groups in D and the exterior $\bar{D}$. This implies that the number of linearly independent closed curves on S^* is double the number of those in D. The three authors, mentioned above, use also the notion of the zero-dimensional homology. Two points are said to be homologous if they bound an arc in D. Let the boundary S^* of D consist of components $S^* = S_0^* + \cdots + S_q^*$. To each component S_k^* one may assign a point P_k in the component of the exterior $\bar{D}$ (on S_0); $\bar{D}$ has as many separated components as the common boundary S^*. The zero-dimensional homology group of $\bar{D}$ is generated by the basis P_0P_k (P_0 on S_0) ($k = 1, \cdots, q$). If P_0 and P_k are chosen on the boundary, then the arcs $C_k(\equiv P_0P_k)$ constitute the basis for the homology (in D) of open curves modulo the boundary. Two curves C and C' in D which initiate and terminate on S^* are homologous modulo the boundary ($C \sim C'$ [mod S^*]) if their zero-dimensional boundaries are homologous in $\bar{D}$. This is equivalent to the set $C \sim C'$ [mod S^*], if $C - C'$ can be completed to a closed curve which bounds in D by adding arcs on the boundary only. Thus it is seen that the homology group of open arcs modulo the boundary in D is isomorphic to the zero-dimensional homology group in the exterior $\bar{D}$.

---

†The subscript "t" denotes the tangential boundary values.

Similarly, the bases S_k^* and $C_k (k = 1, \cdots, q)$ may be used according to the Alexander duality theorem to establish a duality between two homology groups in D: the group of closed surfaces $\{S\}$, and the group of open curves $\{C\}$. The bases S_k^* and C_k are reciprocal: the number of intersections of S_k^* and C_j is equal to 1 if $j = k$ and equal to zero otherwise.

We may briefly discuss the homology of open surfaces modulo the boundary. There are dual bases, Γ_k in D and $\hat{\Gamma}_j$ in $\bar{D}$; Γ_k links $\hat{\Gamma}_j$ exactly once if $j = k$, and not at all if $j \neq k$. By taking the representative curves Γ_k and $\hat{\Gamma}_j$ on the boundary, S, we can obtain in D a duality between the closed curve homology and the homology of open surfaces modulo the boundary. The curves $\hat{\Gamma}_k$ do not bound in $\bar{D}$ or on S^*, but they may bound surfaces Σ_k in D. The intersection number of Γ_j with Σ_k is one if $j = k$ and 0 otherwise (the bases Γ_j and Σ_k are reciprocal). The homology of open surfaces Σ_k is determined by the homology of the boundary closed curves $\hat{\Gamma}_k$. One may find an equivalent definition to the above of the following form: if it is possible to complete $\Sigma_k - \Sigma_k'$ to a bounding closed surface in D by adding a part of S^*, then the above definition is equivalent to setting $\Sigma_k - \Sigma_k' \sim 0$ [mod S^*].

We may discuss some characteristic properties of the homology modulo the boundary. Let us begin with the solenoidal vector Y. The surface integral $\oint_S Y_n \, dS$ depends only upon the homology class alone and not upon the special choice of the surface S. Choosing $\operatorname{div} Y = 0$ and $Y_n = 0$ on S^* assures that the periods $[S;Y]$ vanish on the basis $S_k^* (k = 1, 2, \cdots, q)$ located on S^* and as a consequence all closed surface periods vanish. Then it is possible to define open surface periods by means of the homology [mod S^*]. Let $[\Sigma; Y] = \int_\Sigma Y \cdot dS$, Σ in D with boundary on S^*. Since for any $\Sigma' \sim \Sigma$ [mod S^*], $\Sigma - \Sigma'$ can be completed to a closed bounding surface by adding a part of S^*, the period $[\Sigma; Y]$ is only a function of the homology class alone. Conversely, if $\int_\Sigma Y \cdot dS$ are periods (depend only on the homology class) then the closed surface periods $[S;Y]$ vanish and $Y_n = 0$ on S^*. Define a local surface gradient on S as a vector A such that $\oint_\Gamma A \cdot dx = 0$ for any Γ bounding on S. Then from $\operatorname{div} Y = 0$, $Y = \operatorname{curl} A$, i.e., if the open surface integrals are periods, there exists a local single-valued vector potential $A = \nabla_S \phi$ $[\phi(x) = \int_{x_0}^{x} A \cdot dx$, the path of integration located on $S]$

which is a local surface gradient on S^*. One has

$$\oint_{\Gamma^*} A \cdot dx = \int_{\Sigma^*} \operatorname{curl} A \cdot dS = \int_{\Sigma^*} Y_n \, dS = 0,$$

with Σ^* being a surface element on S^* with Γ^* as its boundary. If all periods $[\Sigma; Y]$ vanish then A is the surface gradient of a single-valued potential on S^*.

Consider an irrotational vector X; the closed curve integral $\oint_\Gamma X \cdot dx$ is a period. With the boundary condition $X_t = 0$ on S^*, the periods $[\Gamma; X]$ vanish on the basis $\Gamma_k^* (k = 1, \cdots, p)$ located on S^*, and hence all closed curve periods vanish. The open curve integrals for C with boundary on S^* are now periods, $[C; X] = \int_C X \cdot dx$. If the integrals $\int_C X \cdot dx$ are periods, then the closed curve periods $[\Gamma; X]$ vanish and $X_t = 0$ on S^*. Since $\operatorname{curl} \operatorname{grad} \phi = 0$, then there exists a local scalar potential ϕ. If the open curve integrals are periods, then ϕ is single-valued in D and constant on each component of S^*. If all values $[C; X]$ vanish, then ϕ takes the same constant value on each component of S^*.

As the next item we shall discuss the compatibility of an irrotational vector. Consider a vector X with $\operatorname{curl} X = 0$ in D and the restrictions which are imposed upon the assignment of tangential boundary values X_t on S^*. For any Γ^* bounding a surface element Σ^* on S^* we have

$$\oint_{\Gamma^*} X \cdot dx = \int_{\Sigma^*} \operatorname{curl} X \cdot dS = 0, \qquad \Gamma^* \sim 0 \text{ on } S^*.$$

It was shown above that if X_t vanishes for each curve Γ^* bounding on S^*, then X is a local surface gradient. On S^*, the homology group of closed curves has as a basis the $2p$ curves: $\Gamma_1^*, \cdots, \Gamma_p^*$ as a basis for the homology group of closed curves in D, and $\hat{\Gamma}_1^*, \cdots, \hat{\Gamma}_p^*$ for the homology group of bound surfaces $\Sigma_1, \cdots, \Sigma_p$ in D; this is a reciprocal basis for the homology group of open surfaces. We have

$$[\hat{\Gamma}_k^*; X] = \oint_{\hat{\Gamma}_k^*} X \cdot dx = \oint_{\hat{\Gamma}_k^*} X_t \cdot dx = \int_{\Sigma_k} \operatorname{curl} X \cdot dS = 0. \tag{3.10}$$

This implies that the boundary values X_t cannot be assigned freely to an irrotational X in D; a half of the closed curve periods of X_t on S^* must vanish; i.e.,

$$[\hat{\Gamma}_k^*;X] = \oint_{\hat{\Gamma}^*} X_t \cdot dx = 0 \qquad (k = 1, \cdots, p).$$

This implies that for an irrotational X to assign the field X_t compatibly on S^*, the tangential field has to be a local surface gradient; a half of its periods must vanish while the remainder are determined by the given boundary values (consider on a torus a helical field X_t on S^*, with X_t being a local surface gradient; from the two fundamental closed curves on S^*, Γ bounds the exterior domain (horizontal cross sections), $\hat{\Gamma}$ the interior (vertical cross sections); for an irrotational X inside the torus the transverse period $[\hat{\Gamma};X]$ must vanish to satisfy the compatibility conditions; to be compatible with $\operatorname{curl} X = 0$ in the exterior of the torus, the longitudinal period $[\Gamma;X]$ must vanish). As the next item let us consider an irrotational vector satisfying the homogeneous boundary conditions, $X_t = 0$, i.e., all closed curve periods vanish and the compatibility conditions are satisfied. Open curve periods are defined by $[C;X] = \int_C X \cdot dx$, C having end points on S^*, with $X = \operatorname{grad} \phi$, ϕ taking constant values ϕ_k on the boundary component S_k^*. The periods on a basis are the potential differences $[C_k;X] = \phi_k - \phi_0$, C_k connecting S_k^* to S_0^*, i.e., for $X_t = 0$, there exist vectors X for each possible specification of the constant values of the potentials on the boundary surfaces S_k^*.

The discussion of the compatibility for the values of the normal boundary component of a solenoidal vector, Y, will be presented very briefly. With $\operatorname{div} Y = 0$ in D, $S^* = S_0^* + \cdots + S_q$, ($S_k^*$ are separate components of the boundary), $S^* \sim 0$ in D, the values of Y_n will be assigned compatibly:

$$\oint_{S^*} Y_n \, dS = \sum_{k=0}^{q} [S_k^*;Y] = 0 \,. \tag{3.11}$$

With the homogeneous B.C., $Y_n = 0$, all closed surface periods of this solenoidal field vanish and the open surface periods are determined in terms of the homology modulo the boundary. With Σ being a surface in D with boundary on S^*, we have

$$[\Sigma; Y] = \int_{\Sigma} Y \cdot dS. \tag{3.12}$$

Thus there always exists a solenoidal Y for every arbitrarily given Y_n preserving (3.11) and in the case $Y_n = 0$, the open surface periods (3.12) may have freely assigned values on a basis $\Sigma_1, \cdots, \Sigma_p$.

The compatibility conditions for boundary value problems with inhomogeneous differential equations $\operatorname{curl} X = \Xi$ ($\operatorname{div} \Xi = 0$) and $\operatorname{div} Y = \sigma$, σ = given and known, are a little more complicated. For $\operatorname{div} Y = \sigma$ we have the compatibility condition [in place of (3.11)]:

$$\oint_{S^*} Y_n \, dS = \int_D \sigma \, dV . \tag{3.13}$$

The integral $\oint_S Y \cdot dS$ is not constant on the homology class of S. But if $S' \sim S$, and D' is domain bounded by $S' - S$, then

$$\oint_{S'} Y \cdot dS - \oint_S Y \cdot dS = \int_{D'} \sigma \, dV . \tag{3.14}$$

Hence, the period is $[S; Y] = \oint_S Y \cdot dS$, but the integral has to be calculated for a specified member of each homology class with the known values of σ (one has to take fixed representatives in each homology class). Consider

$$\operatorname{curl} X = \Xi \qquad (\operatorname{div} \Xi = 0), \tag{3.15}$$

where Ξ is a given vector in D. Again the values of $\oint_\Gamma X \cdot dx$ do not depend only upon the homology class but also on the values of the given vector. With $\Gamma' \sim \Gamma$, $\Gamma' - \Gamma$ bounds a surface Σ in D and

$$\oint_{\Gamma'} X \cdot dx - \oint_\Gamma X \cdot dx = \int_\Sigma \Xi \cdot dS . \tag{3.16}$$

As above, specific values of the periods $[\Gamma; X] = \oint_\Gamma X \cdot dx$ must be assigned to fixed representatives of the homology classes. Concerning the compatibility conditions on the boundary values X_t, the vector X_t is not any longer a local surface gradient; in place of $\oint_{\Gamma^*} X_t \cdot dx = 0$, this vector must satisfy the condition

$$\oint_{\Gamma^*} X_t \cdot dx = \int_{\Sigma^*} \Xi \cdot dS \, , \tag{3.17}$$

where Σ^* is an arbitrary surface domain on S^* with boundary Γ^* (the curl X_t on the surface is a given function on Σ^*; this defines X_t uniquely to within an arbitrarily surface gradient). Half of the closed curve periods on S^* are defined by the values of Ξ; instead of the condition

$$[\hat{\Gamma}_k^*; X] = \oint_{\hat{\Gamma}_k^*} X_t \cdot dx = 0 \, ,$$

one has to use the condition†

$$\oint_{\hat{\Gamma}^*} X_t \cdot dx = \int_{\Sigma} \Xi \cdot dS, \qquad \Sigma \equiv \Sigma_k \, ;$$

this period $[\hat{\Gamma}_k^*; X]$ depends now upon the choice of the homology class. Briefly, the situation may be presented in the following way: Let X^0 be any vector satisfying $\operatorname{curl} X^0 = \Xi$; it has certain values X_t^0 on S^* and certain periods $\oint_{\hat{\Gamma}_k^*} X_t^0 \cdot dx$ on specified $\hat{\Gamma}_k^*$. For some boundary values X_t to be compatible with given Ξ, X_t may differ from X_t^0 by a local surface gradient and has to have periods equal to those of X_t^0 on $\hat{\Gamma}_k^*$.

A few remarks may be given on the concept of a period on open curves and surfaces for inhomogeneous boundary conditions. With $\operatorname{curl} X = 0$, X_t given, the integral $\int_C X_t \cdot dx$ is not a function of homology class [mod S^*] alone and the values of the integral for two curves in a homology class differ by the value of a curvilinear integral on S^* which is known from the values of X_t. With $\operatorname{div} Y = 0$, Y_n given, the surface integrals on two open surfaces in a homology class differ by an integral on a surface domain on S^* which is known from the values of Y_n.

† $\hat{\Gamma}^* = \hat{\Gamma}_k^*$.

4. Some Existence Theorems

Blank *et al.* ([1], p. 54 ff.) attack the problem of the existence theorems in the following way:

The solution of the systems of equations derived in Section 2, i.e., Eqs. (2.14) to (2.17) and (2.18), (2.19) reduces to the solution of the system (2.23). The solution of (2.23) is treated as the sum of solutions of three systems of equations:

$$\operatorname{div} Z = 0; \qquad \operatorname{curl} Z = 0; \tag{4.1}$$

$$\operatorname{div} Z = \sigma; \qquad \operatorname{curl} Z = 0; \tag{4.2}$$

$$\operatorname{div} Z = 0; \qquad \operatorname{curl} Z = \Xi; \qquad \operatorname{div} \Xi = 0. \tag{4.3}$$

The solutions of the system (4.1) are harmonic vectors. The particular solution of the system (4.2) is

$$Z = \operatorname{grad} \phi; \qquad \phi = -(4\pi)^{-1} \int_D \sigma r^{-1}\, dV, \tag{4.4}$$

and of the system (4.3)

$$Z = \operatorname{curl} A; \qquad A = (4\pi)^{-1} \int_D r^{-1} \Xi\, dV. \tag{4.5}$$

A solution of the system (2.23) or (3.1) is

$$Z = H + \operatorname{grad} \phi + \operatorname{curl} A, \tag{4.6}$$

where H is the harmonic vector which has to be defined uniquely through the properly posed boundary conditions (any vector can be presented as the sum of a harmonic vector, a gradient and a curl). Consider harmonic vectors in D. For D being a simply connected, Z may be chosen in the form $Z = \nabla\phi$ $(\Delta\phi = 0)$. Hence the question of selecting the harmonic Z reduces to the solution of the scalar Laplace equation with ϕ or

$$\frac{\partial\phi}{\partial n}\left(\oint \frac{\partial\phi}{\partial n}\, dS = 0\right)$$

given on the boundary; analogously, the boundary conditions on Z may be given in the form of Z_t as an arbitrary surface gradient or Z_n with $\oint Z_n \, dS = 0$. A harmonic Z has well-defined periods

$$\oint Z \cdot dx = [\Gamma] \,, \qquad \oint Z \cdot dS = [S] \,.$$

For $Z_n = 0$, the periods $[\Sigma] = \int Z \cdot dS$ are meaningful, the periods $[S]$ vanish. For $Z_t = 0$, the periods $[C] = \int Z \cdot dx$ are meaningful, all $[\Gamma]$ vanish. In general, the specific periods $[\Gamma]$ and $[\Sigma]$ are not equal in value; there may be a correspondence between them in an isomorphism of $\{\Sigma\}$ and $\{\Gamma\}$.

Hence we have four homology groups in D: closed surface $\{S\}$, open curve $\{C\}$, closed curve $\{\Gamma\}$, open surface $\{\Sigma\}$. The reciprocal bases exhibit a duality between $\{\Gamma\}$ and $\{\Sigma\}$, and $\{S\}$ and $\{C\}$ (i.e., there are two abstract groups). The (C,S) or $\mathcal{E}$ group is associated with the electric vector; its periods may be associated with emf, $V = [C;E]$, or with charges, $Q = \kappa[S;E]$. The (Γ,Σ) or $\mathfrak{m}$ group refers to the magnetic vector; its periods are defined by flux $\Phi = [\Sigma;B]$ and current $\mu I = [\hat{\Gamma};B]$ in external circuits. A well-posed problem gives boundary values of Z_n and the values of a set of $\mathfrak{m}$ periods; or of Z_t and a set of $\mathcal{E}$ periods. The values of the dual periods $[\Sigma]$ and $[\Gamma]$ are not necessarily equal but they are related so that specification of one (jointly with the boundary conditions) determines the other; hence only one set of periods for the $\mathfrak{m}$ group can be given.

Below we cite some theorems of Blank, Friedrichs, and Grad.

THEOREM 1. There exists a unique harmonic vector Z with arbitrarily assigned $\mathfrak{m}$ periods (in either the $\{\Sigma\}$ or $\{\Gamma\}$ representation) and with arbitrarily assigned values Z_n on the boundary S^* subject to the compatibility condition (3.13),

$$\oint_{S^*} Z_n \, dS = [S^*;Z] = 0 \,.$$

Corollary. There exists a unique harmonic vector Z, with arbitrarily specified $\mathfrak{m}$ periods for which $Z_n = 0$ on S^*.

THEOREM 2. There exists a unique harmonic vector with arbitrarily assigned $\mathcal{E}$ periods in either the $\{S\}$ or $\{C\}$ representation and with boundary values Z_t which are local surface gradients and satisfy the compatibility condition

$$[\hat{\Gamma}_k^*; Z] = \oint_{\hat{\Gamma}_k^*} Z_t \cdot dx = 0 \quad \text{on} \quad S^*,$$

but are otherwise arbitrary.

Corollary. There exists a unique harmonic vector with arbitrary assigned $\mathcal{E}$ periods in either representation and homogeneous boundary values $Z_t = 0$ on S^*.

Concerning theorems for the inhomogeneous differential systems, the following will be cited:

THEOREM 3. The inhomogeneous systems (4.2) and (4.3) possess unique solutions under the conditions of Theorem 1 or 2 with the compatibility conditions corresponding to the prescribed fields σ and Ξ.

THEOREM 4. There exists a unique solution of the system:

$$\text{div}\, A = \sigma; \qquad (\text{div}\, \Xi = 0); \qquad \text{curl curl}\, A = \Xi, \tag{4.7}$$

with arbitrary boundary values A_t and arbitrarily prescribed periods $[C;A]$ (or $[S;A]$) compatible with σ.

5. Application to the System of Electromagnetic Equations

The theorems, derived above, are applied below to the electromagnetic equations (see Blank *et al.* [1], p. 113 ff.). The system (2.18), (2.19) is transformed into the form (with $\dot{B} = \partial B/\partial t$):

$$\text{curl}\, E = -\dot{B}; \qquad \text{div}\, E = \kappa^{-1} q; \tag{5.1}$$

$$\text{curl}\, \dot{B} = \mu \dot{j}; \qquad \text{div}\, \dot{B} = 0, \tag{5.2}$$

with $\dot{j}$ and q assumed to be given as functions of space coordinates at a given time t. With $\text{curl curl}\, E = -\mu \dot{j}$, we get from Theorem 4:

THEOREM 5. There exists a unique solution of (5.1), (5.2) for E and $\dot{B}$ possessing arbitrary boundary values E_t and $\mathcal{E}$ periods in closed surface representation $[S;E]$, compatible with q or open curve $[C;E]$.

Blank, Friedrichs, and Grad show that E is given by minimizing the expression

$$F[E] = \int_D \frac{1}{2} |\operatorname{curl} E|^2 dV + \int_D E \cdot \mu \dot{J} dV, \tag{5.3}$$

subject to the admissibility condition

$$\operatorname{div} E = \kappa^{-1} q; \quad E_t \text{ given on } S^*; \quad [S;E] \text{ or } [C;E] \text{ given.} \tag{5.4}$$

Obviously $\dot{B} = -\operatorname{curl} E$. Another solution is obtained by minimizing

$$F[\dot{B}] = \int_D \frac{1}{2} \dot{B}^2 dV + \oint_{S^*} E_t \times \dot{B} \cdot dS, \tag{5.5}$$

subject to the admissibility condition

$$\operatorname{curl} \dot{B} = \mu \dot{J}. \tag{5.6}$$

As it is seen only the given and known boundary values, E_t, appear in Eq. (5.5) (well-defined problem).

One may get the second existence theorem from Theorem 3 (at first to $\dot{B}$ and next to E):

THEOREM 6. A unique solution of (5.1), (5.2) exists with given boundary values $\dot{B}_n$ and $\mathfrak{m}$ periods ($[\Sigma;\dot{B}]$ or $[\Gamma;\dot{B}]$) as well as given boundary values E_n and $\mathfrak{m}$ periods ($[\Sigma;E]$ or $[\Gamma;E]$).

One may also construct well-posed problem and theorems for E and B (instead of for E and $\dot{B}$) (see Blank *et al.* [1], p. 119).

In the approach accepted in the present work, each of the theorems, presented above, is valid independently in each step of the chosen successive approximation procedure.

6. Application to the System of Hydrodynamic Equations

As a new feature of the application of the theory outlined above, we shall apply some of the theorems, presented above, to the system of hydrodynamic equations. Consider the system of equations (2.16), (2.17). In accordance with the remarks given above, the right-hand sides of these equations are supposed to be known and given in each step of the successive approximation procedure. A solution for q in the subsystem of Eqs. (2.16), (2.17) furnishes the vector q. The remaining functions a^2, p, ρ, T, can be found easily from the remaining equations. In general, in each step of the successive approximation procedure this subsystem can be represented by means of a system of the form

$$\operatorname{curl} M = N; \qquad \operatorname{div} M = R; \tag{6.1}$$

$$M \equiv q; \qquad N \equiv Z; \qquad R \equiv W. \tag{6.2}$$

We may apply Theorems 1 and 3 to M (i.e., to q) and obtain:

THEOREM 7. A unique solution of (6.1) [or of (2.16), (2.17)] exists with given boundary values M_n (i.e., $A_n = q_n$) and $\mathfrak{m}$ periods: open surface periods $[\Sigma;A]$, or close curve periods $[\Gamma;A]$, compatible with N.

We may discuss the period conditions. For the vector field of the velocity q, the $\mathfrak{m}$ period $[\Sigma;A]$ may be generally assumed to be relevant to the physical applications and notions. It may be referred to as the total (velocity) flux concept through the surface Σ. The period $[\Gamma;A]$ can be interpreted as the circulation.

We may investigate the possibility of using the tangential components. Using the Theorems 2 and 3, we may put down:

THEOREM 8. A unique solution of (6.1) [or of (2.16), (2.17)] exists with given boundary values M_t (i.e., $A_t = q_t$) and $\mathcal{E}$ periods in either representations: open curve periods $[C;A]$ or closed surface periods $[S;A]$ compatible with R.

The physical interpretation of the periods in this case is more complicated, if in general possible.

Let us remodel the system (6.1) in the sense

$$\operatorname{curl}\operatorname{curl} M = \Xi; \qquad \operatorname{div} M = R; \qquad \operatorname{div} \Xi = 0, \tag{6.3}$$

which can be treated as a system consisting of two subsystems

$$\operatorname{curl} Z = \Xi; \qquad \operatorname{div} Z = 0; \qquad (\operatorname{div} \Xi = 0); \tag{6.4}$$

$$\operatorname{curl} M = Z; \qquad \operatorname{div} M = R; \qquad (\operatorname{div} Z = 0). \tag{6.5}$$

In this case we can apply Theorem 4:

THEOREM 9. There exists a unique solution of the system (6.3) with arbitrary boundary values M_t (i.e., $A_t = q_t$) and arbitrarity prescribed periods $[C;M]$ (or $[S;M]$) compatible with R.

In a similar way we may generalize other theorems.

7. Successive Approximation Procedure

Above, we assumed that the right-hand sides of equations in question are known and given, which requires some sort of a successive approximation procedure. Denoting the particular steps in this procedure by the letter "m", we have the following system of equations:

(i) hydrodynamical:

$$\operatorname{div}(q)_{m+1} = -\left[\rho^{-1}(\operatorname{grad}\rho \cdot q + \rho_{,t})\right]_m ; \tag{7.1}$$

$$\operatorname{curl}(q)_{m+1} = q_m^{-2}\left\{(q\cdot\omega)q - q \times \left[q_{,t} + \nabla(H-Q) - P\right]\right\}_m ; \tag{7.2}$$

(ii) electromagnetic:

$$\operatorname{curl} B_{m+1} = \mu J_m ; \tag{7.3a}$$

$$\operatorname{div} B_{m+1} = 0 ; \tag{7.3b}$$

$$\operatorname{curl} E_{m+1} = -\partial B_m/\partial t ; \tag{7.4a}$$

$$\operatorname{div} E_{m+1} = \kappa^{-1}\alpha . \tag{7.4b}$$

The quantities ω_m, J_m, Q_{1m}, P_{1m} appearing in the system of equations are calculated directly from the following equations:

$$\operatorname{curl} q_m = \omega_m ; \tag{7.5}$$

$$J_m = \sigma_1(E_m + \mu q_m \times H_m) ; \tag{7.6}$$

$$Q_{1m} = J_m^2\sigma_1^{-1} ; \tag{7.7}$$

$$P_{1m} = \alpha E_m + J_m \times B_m ; \tag{7.8}$$

moreover, the velocity of sound, a^2 (and next the pressure p and the temperature T) can be calculated from Eq. (2.13). Thus, the expressions (7.5) to (7.8) do not need to be considered. There remains the question of the convergence of the successive approximation procedure.

The systems (7.1), (7.2), (7.3), and (7.4) are of the form

$$\operatorname{curl} A_{m+1} = N_m ; \qquad \operatorname{div} A_{m+1} = \sigma_m , \tag{7.9}$$

or of the form

$$\operatorname{curl}\operatorname{curl} A_{m+1} = \operatorname{curl} N_m = S_m ; \qquad \operatorname{div} A_{m+1} = \sigma_m . \tag{7.10}$$

Blank *et al.* propose a variational solution of the system [see Eq. (5.3)]:

$$\operatorname{curl}\operatorname{curl} A = S ; \qquad \operatorname{div} A = \sigma , \tag{7.11}$$

admissible: $\operatorname{div} A = \sigma ;$ (7.11a)

variational: $\operatorname{curl}\operatorname{curl} A = S ,$ (7.11b)

which system admits an arbitrary specification of the boundary values A_t with the given periods $[S;A]$ or $[C;A]$. We propose that the vector A_{m+1} is then given by minimizing the expression

$$F[A_{m+1}] = \int_D \frac{1}{2} \left| \operatorname{curl} A_{m+1} \right|^2 dV - \int_D A_{m+1} \cdot S_m \, dV , \tag{7.12}$$

which in the case of the present system and of the applied successive approximation procedure takes the form

$$F[q_{m+1}] = \int_D \frac{1}{2} \left| \operatorname{curl} (q_{m+1}) \right|^2 dV - \int_D (q_{m+1}) \cdot S_m \, dV ; \tag{7.13}$$

$$S_m = q_m^{-2} \left\{ (q \cdot \omega) q - q \times \left[q_{,t} + \nabla (H - Q) - P \right] \right\}_m ; \tag{7.14}$$

$$F[B_{m+1}] = \int_D \frac{1}{2}|\text{curl } B_{m+1}|^2 \, dV - \int_D B_{m+1} \cdot S_m^{(10)} \, dV ; \tag{7.15}$$

$$F[E_{m+1}] = \int_D \frac{1}{2}|\text{curl } E_{m+1}|^2 \, dV - \int_D E_{m+1} \cdot S_m^{(20)} \, dV ,$$

$$m = 1, 2, 3, \cdots . \tag{7.16}$$

As mentioned above, the tangential values of the functions in question on the boundary,

$$q_t; \quad B_t; \quad E_t, \tag{7.17}$$

are arbitrary and given. The zero-approximation values

$$\rho_0(x;t); \quad q_0(x;t); \quad \omega_0(x;t); \quad B_0(x;t); \quad E_0(x;t), \tag{7.18}$$

are also known and given.

The tool of the calculus of variation which has to be applied to minimize the expression (7.12) is a well-known one and does not need to be repeated here. What we may attempt to demonstrate is the fact that it should be possible to show that the sequence of functions A_{m+1}, obtained in the manner described above, converges uniformly to a function A, under some restrictive conditions, presented below.* For the simplicity sake we concentrate only upon the simple variational formulation, referring to Eq. (7.12), neglecting admissible conditions like those given by Eq. (7.11a). Let us use the symbol F for the expression [see Eq. (7.12)]

$$F = \frac{1}{2}|\text{curl } A_{m+1}|^2 - A_{m+1} \cdot S_m , \tag{7.19}$$

where the vectors A_{m+1} and S_m have the components

*The author emphasizes very strongly that the discussion, presented below, does not provide any actual proof of the convergence. It is only a demonstration that it should be possible in some cases to construct such a proof.

$$A_{m+1} = \left(A_{m+1}^{(1)},\ A_{m+1}^{(2)},\ A_{m+1}^{(3)}\right); \qquad S_m = \left(S_m^{(1)},\ S_m^{(2)},\ S_m^{(3)}\right). \tag{7.20}$$

The Euler-Lagrange equations are*:

$$\frac{\partial F}{\partial w} - \frac{\partial}{\partial x}\left(\frac{\partial F}{\partial w_{,x}}\right) - \frac{\partial}{\partial y}\left(\frac{\partial F}{\partial w_{,y}}\right) - \frac{\partial}{\partial z}\left(\frac{\partial F}{\partial w_{,z}}\right) = 0, \tag{7.21}$$

with

$$w = A_{m+1}^{(i)}, \qquad i = 1, 2, 3. \tag{7.22}$$

The entire reasoning presented below is of a very general character. We may neglect the absolute values in Eq. (7.19); the general character of our discussion is not affected by this item and the reasoning will be valid in this and other similar cases. In particular,

$$\begin{aligned} F = \frac{1}{2}\Bigg\{ &\left(\frac{\partial A_{m+1}^{(3)}}{\partial y}\right)^2 - 2\,\frac{\partial A_{m+1}^{(3)}}{\partial y}\,\frac{\partial A_{m+1}^{(2)}}{\partial z} + \left(\frac{\partial A_{m+1}^{(2)}}{\partial z}\right)^2 \\ &+ \left(\frac{\partial A_{m+1}^{(1)}}{\partial z}\right)^2 - 2\,\frac{\partial A_{m+1}^{(1)}}{\partial z}\,\frac{\partial A_{m+1}^{(3)}}{\partial x} + \left(\frac{\partial A_{m+1}^{(3)}}{\partial x}\right)^2 \\ &+ \left(\frac{\partial A_{m+1}^{(2)}}{\partial x}\right)^2 - 2\,\frac{\partial A_{m+1}^{(2)}}{\partial x}\,\frac{\partial A_{m+1}^{(1)}}{\partial y} + \left(\frac{\partial A_{m+1}^{(1)}}{\partial y}\right)^2 \\ &- A_{m+1}^{(1)} S_m^{(1)} - A_{m+1}^{(2)} S_m^{(2)} - A_{m+1}^{(3)} S_m^{(3)} \Bigg\}, \end{aligned} \tag{7.23}$$

and for $w = A_{m+1}^{(1)}$ we obtain from Eq. (7.21):

$$\left(A_{m+1,yy}^{(1)} + A_{m+1,zz}^{(1)}\right) - A_{m+1,xy}^{(2)} - A_{m+1,xz}^{(3)} + S_m^{(1)} = 0, \tag{7.24}$$

*The variational problem is treated as that of determining extreme of multiple integrals in several dependent (unknown) functions and several independent variables.

with the analogous equations for $A^{(2)}_{m+1}$ and $A^{(3)}_{m+1}$. This is a system of three interrelated equations for the three unknown functions $A^{(i)}_{m+1}$, $i = 1, 2, 3$, which can be solved by some sort of a successive approximation process or of iteration process in which the terms containing the functions $A^{(2)}_{m+1}$ and $A^{(3)}_{m+1}$ are transferred on the right-hand side of Eq. (7.24). Let us add on both sides of Eq. (7.24) the term $A^{(1)}_{m+1,xx}$ and in the next step let us assume that the right-hand side term is approximated by $A^{(1)}_{m,xx}$ in place of the term $A^{(1)}_{m+1,xx}$. This implies that Eq. (7.24) can be approximated by the equation

$$\nabla^2 A^{(1)}_{m+1} = (\operatorname{div} A_m)_{,x} - S^{(1)}_m , \tag{7.25}$$

with the analogous equations for $A^{(i)}_{m+1,jj}$, $i = 2, 3$. Let us assume that Eq. (7.25) may be solved by means of some sort of an integral operator method; hence let us assume that we can write for the illustrative purposes and as an example

$$A^{(1)}_{m+1} = \int_{D'} K^{(1)}(x, y, z; \xi, \eta, \zeta; t)\, H^{(1)}_m(\xi, \eta, \zeta; t)\, dV' , \tag{7.26}$$

where

$$H^{(1)}_m(x, y, z; t) = \left[\operatorname{div} A_m(x, y, z; t)\right]_{,x} - S^{(1)}_m(x, y, z; t) ,$$

$$\int_{D'} \cdots dV' = \int_0^{\xi} \cdots d\xi_1 \int_0^{\eta} \cdots d\eta_1 \int_0^{\zeta} \cdots d\zeta_1 , \tag{7.27}$$

and the function $K^{(1)}(x, y, z; \xi, \eta, \zeta; t)$ is a kernel function. We assume that the functions $A^{(i)}_0$ $(i = 1, 2, 3)$ and all their partial derivatives up to the second order ($A^{(i)}_{0,x}$; $A^{(i)}_{0,xy}$; $A^{(i)}_{0,zx}$; $A^{(i)}_{0,t}$; etc.) are known, continuous and bounded functions in the domain in question:

$$\left|A^{(i)}_0\right|; \quad \left|A^{(i)}_{0,x}\right|; \quad \left|A^{(i)}_{0,y}\right|; \quad \left|A^{(i)}_{0,xy}\right|; \quad \left|A^{(i)}_{0,zx}\right|; \quad \text{etc.}, \leqq c_0 . \tag{7.28}$$

Concerning the function $K^{(1)}$ we assume that $K^{(1)}(x,y,z;\xi,\eta,\zeta)$ and its all partial derivatives up to the second order are well-behaving functions in D or D', except possibly at a finite number of points. Hence we assume that, with the exception of a finite number of points, the function $K^{(1)}$ and all its partial derivatives with respect to (x,y,z) up to the second order and the first derivative with respect to t are bounded, i.e.,

$$\left|K^{(1)}\right|; \quad \left|K^{(1)}_{,x}\right|; \quad \left|K^{(1)}_{,y}\right|; \cdots ; \quad \left|K^{(1)}_{,xx}\right|; \quad \left|K^{(1)}_{,yy}\right|; \cdots , \leqq c_1. \tag{7.29}$$

Differentiating the expression (7.26) for $m=0$ with respect to $(x,y,z;\ t)$ with D' being finite and with the inequalities (7.28) taken into account furnishes the first- and second-order partial derivatives of $A_1^{(1)}$ which are bounded, except possibly at a finite number of points, i.e.,

$$\left|A^{(1)}_{1,xx}\right|; \quad \left|A^{(1)}_{1,xy}\right|; \cdots , \qquad \text{etc.}, \leqq \bar{c}_0 . \tag{7.29a}$$

By induction, the inequalities (7.29) are valid for $m \geqq 1$. Without a loss of generality we can take $\bar{c}_0 \leqq c_0$. We consider three sequences of functions:

(i) the sequence of functions:

$$A_0^{(1)} = G_0 ; \tag{7.30}$$

$$\vdots \qquad \vdots$$

$$A^{(1)}_{m+1} = \int_{D'} K^{(1)} H^{(1)}_m \, dV' ; \tag{7.31}$$

we wish to show that the sequence defined by (i) converges uniformly (except possibly at a finite number of points) to a function $A^{(1)}$ for $0 \leqq D' \leqq V'$; if so, we may pass to the limit under the sign of integration in (i) as $m \to \infty$, obtaining

$$A^{(1)} = \int_{D'} K^{(1)} (x,y,z;\xi,\eta,\zeta;\ t) \Big\{ \big[\operatorname{div} A(\xi,\eta,\zeta;\ t)\big]_{,\xi}$$

$$- S^{(1)} (\xi,\eta,\zeta;\ t) \Big\} \, dV' ; \tag{7.32}$$

(ii) the sequence of the first partial derivatives:

$$A^{(1)}_{0,j} = G_j \qquad (j = x, y, z); \tag{7.33}$$

$$\vdots \qquad \vdots$$

$$A^{(1)}_{m+1,j} = \int_{D'} K^{(1)}_{,j} \, H^{(1)}_m \, dV; \tag{7.34}$$

as above, we obtain for $m \to \infty$:

$$A^{(1)}_{,j} = \int_{D'} K^{(1)}_{,j} \, H^{(1)} \, dV'; \tag{7.35}$$

(iii) the sequence of the second partial derivatives:

$$A^{(1)}_{0,ij} = G_{ij} \qquad (i, j = x, y, z); \tag{7.36}$$

$$\cdots\cdots\cdots\cdots\cdots\cdots$$

$$A^{(1)}_{m+1,ij} = \int_{D'} K^{(1)}_{,ij} \, H^{(1)}_m dV', \tag{7.37}$$

with the analogous results, as above.

To demonstrate the existence of the convergence of the three sequences, {i}, {ii}, {iii}, we consider the series (with x standing for x, y, z):

$$s^{(1)}_1(x,t) = \sum_{m=0}^{\infty} \left(A^{(1)}_{m+1} - A^{(1)}_m \right); \tag{7.38}$$

$$s^{(1)}_2(x,t) = \sum_{m=0}^{\infty} \left(A^{(1)}_{m+1,j} - A^{(1)}_{m,j} \right); \tag{7.39}$$

$$s^{(1)}_3(x,t) = \sum_{m=0}^{\infty} \left(A^{(1)}_{m+1,ij} - A^{(1)}_{m,ij} \right). \tag{7.40}$$

The nth partial sums, $s^{(1)}_{iN}$ (i = 1, 2, 3),

$$s_{1N}^{(1)} = \sum_{m=0}^{N}\left(A_{m+1}^{(1)} - A_{m}^{(1)}\right); \qquad s_{2N}^{(1)} = \sum_{m=0}^{\infty}\left(A_{m+1,j}^{(1)} - A_{m,j}^{(1)}\right), \quad (7.41)$$

etc., have the simple forms

$$s_{1N}^{(1)} = A_{N+1}^{(1)} - A_{0}^{(1)}; \qquad s_{2N}^{(1)} = A_{N+1,j}^{(1)} - A_{0,j}^{(1)};$$

$$s_{3N}^{(1)} = A_{N+1,ij}^{(1)} - A_{0,ij}^{(1)}. \quad (7.42)$$

Consequently, the series converge uniformly if and only if the sequences converge uniformly. The series will converge uniformly if the scalar series

$$\sum_{m=0}^{\infty}\left\|A_{m+1}^{(1)} - A_{m}^{(1)}\right\|; \qquad \sum_{m=0}^{\infty}\left\|A_{m+1,j}^{(1)} - A_{m,j}^{(1)}\right\|;$$

$$\sum_{m=0}^{\infty}\left\|A_{m+1,ij}^{(1)} - A_{m,ij}^{(1)}\right\| \quad (7.43)$$

converge uniformly. From the recurrence relations of (i), (ii), (iii), we obtain

$$A_{m+1}^{(1)} - A_{m}^{(1)} = \int_{D'} K^{(1)}\left(H_{m}^{(1)} - H_{m-1}^{(1)}\right) dV'; \quad (7.44)$$

$$A_{m+1,j}^{(1)} - A_{m,j}^{(1)} = \int_{D'} K_{,j}^{(1)}\left(H_{m}^{(1)} - H_{m-1}^{(1)}\right) dV', \quad (7.45)$$

$$m \geqq 1, \text{ etc.}$$

Thus

$$\left\|A_{m+1}^{(1)} - A_{m}^{(1)}\right\| \leqq \int_{D'}\left\|K^{(1)}\right\| \cdot \left\|H_{m}^{(1)} - H_{m-1}^{(1)}\right\| dV', \quad m \geqq 1, \quad (7.46)$$

$$\left\|A_{m+1,j}^{(1)} - A_{m,j}^{(1)}\right\| \leqq \int_{D'}\left\|K_{,j}^{(1)}\right\| \cdot \left\|H_{m}^{(1)} - H_{m-1}^{(1)}\right\| dV', \quad m \geqq 1, \quad (7.47)$$

etc., or

$$\left\| A_{m+1}^{(1)} - A_m^{(1)} \right\| \leqq \int_{D'} \left\| K^{(1)} \right\| \cdot \left\| \left(A_m^{(1)} - A_{m-1}^{(1)} \right)_{,xx} \right.$$

$$\left. + \left(A_m^{(2)} - A_{m-1}^{(2)} \right)_{,yx} + \left(A_m^{(3)} - A_{m-1}^{(3)} \right)_{,zx} - \left(S_m^{(1)} - S_{m-1}^{(1)} \right) \right\| dV', \quad (7.48)$$

with the analogous two inequalities for the sequences (ii) and (iii). With

$$S_m^{(1)} = A_{m,xy}^{(2)} + A_{m,xz}^{(3)} - A_{m,yy}^{(1)} - A_{m,zz}^{(1)}, \quad (7.49)$$

[see Eqs. (7.9), (7.10)], we get for $m = 0$

$$\left\| A_1^{(1)} - A_0^{(1)} \right\| \leqq \int_{D'} \| 'c_1 \| \cdot \| 7c_0 \| \, dV' \leqq 7 \| 'c_0 \| \cdot \| 'c_1 \| \, X, \quad (7.50)$$

with $X = xyz$. Obviously

$$S_m^{(1)} - S_{m-1}^{(1)} = \left(A_m^{(2)} - A_{m-1}^{(2)} \right)_{,xy} + \left(A_m^{(3)} - A_{m-1}^{(3)} \right)_{,xz}$$

$$- \left(A_m^{(1)} - A_{m-1}^{(1)} \right)_{,yy} - \left(A_m^{(1)} - A_{m-1}^{(1)} \right)_{,zz}, \quad (7.51)$$

which furnishes for $m = 1$

$$\left\| A_2^{(1)} - A_1^{(1)} \right\| \leqq \int_{D'} \| 'c_1 \| \cdot \left\| \left(A_1^{(1)} - A_0^{(1)} \right)_{,xx} \right.$$

$$\left. + \left(A_1^{(1)} - A_0^{(1)} \right)_{,yx} + \cdots - \left(S_1^{(1)} - S_0^{(1)} \right) \right\| dV', \quad (7.52)$$

or

$$\left\| A_2^{(1)} - A_1^{(1)} \right\| \leqq \int_{D'} \| 'c_1 \| \cdot \| 7 \cdot 7 \, 'c_0 \, 'c_1 \| \cdot X \cdot dV'$$

$$\leqq \| 'c_0 \| \, \frac{(7 \, 'c_1 X)^2}{(2!)^3}, \quad (7.53)$$

which leads inductively to the formula

$$\left\| A_{m+1}^{(1)} - A_{m}^{(1)} \right\| \leqq \|c_0\| \cdot \frac{(7'c_1 X)^{m+1}}{\left[(m+1)!\right]^3}, \tag{7.54}$$

for $m = 0, 1, 2, \ldots$, with the analogous inequalities for the derivatives of the first and second order.

The uniform convergence of the exponential series in any finite domain ensures the uniform convergences of all

$$\sum \left\| A_{m+1}^{(1)} - A_{m}^{(1)} \right\|; \quad \left\| A_{m+1,i}^{(1)} - A_{m,i}^{(1)} \right\|;$$

$$\sum \left\| A_{m+1,ij}^{(1)} - A_{m,ij}^{(1)} \right\|,$$

and therefore that of the sequences (i), (ii), (iii), except possibly at a finite number of points. This justifies the mathematical operations performed in previous chapters. It seems that it should be possible to apply the technique of proving the convergence of the successive approximation procedure to methods of solution of Eq. (7.21) other than that chosen in the present chapter. The series (7.54) is of the structure of an exponential series.

Final Remarks

The author generalized the topological technique for the proofs of the existence of the solution, proposed in the past by Blank, Friedrichs and Grad for the pre-Maxwell system, to the system of equations of electromagnetohydrodynamics. Due to the interdependence of the two subsystems involved, electromagnetic and hydrodynamic (this one is nonlinear), the right-hand sides of the equations do contain the unknown variables.

A general discussion on the subject of the successive approximation procedure and the possibility of proving its convergence closes the work.

REFERENCES

[1] A. A. Blank, K. O. Friedrichs, and H. Grad, Notes on Magneto-Hydrodynamics, V. Theory of Maxwell's equations without displacement current. Physics and Mathematics, NYO-6486, AEC Computing and Applied Mathematics Center; Institute of Mathematical Sciences, New York University (Nov. 1, 1957).

[2] G. A. Bliss, The problem of Mayer with variable end-points. *Trans. Am. Math. Soc.* 19, 305-314 (1918).

[3] G. A. Bliss, "Lectures on The Calculus of Variations." Univ. of Chicago Press, Chicago, 1946.

[4] I. Carstoiu, Vorticity and deformation in fluid mechanics. *J. Ratl. Mech. Anal.* 3, 691-712 (1954).

[5] R. von Mises, "Mathematical Theory of Compressible Fluid Flow." Academic Press, New York, 1958.

[6] C. Truesdell. (1) Généralisation de la formule de Cauchy et des théorèmes de Helmholtz au mouvement d'un milieu continu quelconque. *Compt. rend. acad. Sci.* 227, 757-759 (1948).
(2) Une formule pour le vecteur tourbillon d'un fluide visqueux élastique. *ibid.* 821-823.
(3) On the total vorticity of motion of a continuous medium. *Phys. Rev.* 73, 510-512 (1948).
(4) The effect of viscosity on circulation. *J. Meteorol.* 6, 61-62 (1949).
(5) A new vorticity theorem. *Proc. Intern. Congr. Mathematicians* I (1950).
(6) On Ertel's Vorticity Theorem. *Z. angew. Math. Phys.* 2, 109-114 (1951).
(7) Vorticity averages. *Can. J. Math.* 3, 69-86 (1951).
(8) Two measures of vorticity. *J. Ratl. Mech. Anal.* 2, 173-217 (1953).

SOME NEW ASPECTS OF MAGNETO-HYDRODYNAMIC PHENOMENA*

(with Special Emphasis on Magnetohydrodynamic Waves)

John Carstoiu

Science for Industry, Inc.
Brookline, Massachusetts

Introduction

The material presented in this study is not a systematic account of the theory of hydromagnetic wave propagation; it merely summarizes the results of researches undertaken by the writer on this subject in the last year or so, and as such it may rather be considered as an introduction to this theory. It is intended to cast further light on this fascinating subject.

To the pioneer efforts of Alfvén, the discoverer of these waves, Walén and Cowling, and other savants have added their researches resulting in an elegant theory such as is presented in the classical studies of Baños, Grad, Lighthill, and MacDonald (see References).

*Some of our results have already been published in papers listed in the References at the end of this work. Also, some material has been presented in lectures given at New York University, the Institute of Mathematical Sciences, Magneto-Fluid Dynamics Seminar, March 8, 1961, and at the University of Paris, Faculty of Sciences, June 6, 1961. It is a source of great regret that Joseph Pérès who, as the Dean of the Faculty of Sciences, Paris, signed the announcement of my lecture, did not have an opportunity to see this work published before his death.

The subject of magnetohydrodynamics (magnetoaerodynamics, magnetodynamics of conducting fluids or hydromagnetics) is a new one. It is a rapidly advancing subject; its results are not generally known, and indeed some of them are subjects of controversy. Magnetohydrodynamics is concerned with the interaction between magnetic fields and the motion of electrically conducting fluids. The electrical conductivity of the fluid and the prevalence of magnetic fields contribute to the effects of two phenomena: (i) electric currents, induced in the fluid as a result of its motion across the magnetic lines of force, change the existing fields; (ii) the fluid elements carrying currents transverse magnetic lines of force and produce additional forces which modify the motion. Magnetohydrodynamics owes its peculiar interest and difficulty to this twofold interaction between the field and the motion. As Elsasser [23] remarked, "We may describe this roughly as going from a three-dimensional vector space to a six-dimensional vector space. Ordinary hydrodynamics uses the vector space of the velocity vector; the other vector space is the one in which the electromagnetic field vectors are defined. Now, turbulence probably does not exist in two dimensions, but it does exist in three dimensions. What happens in a six-dimensional vector space is anybody's guess."

It is interesting to note that Maxwell had the material to develop the subject to a considerable extent, but only recently has it been attacked significantly. One may also notice that in contrast to the subjects of hydrodynamics and electrodynamics, which now have a well-developed experimental background, relatively little is known experimentally in magnetohydrodynamics. The limitations of the experiments in both space and time are such that they cannot do more than suggest what may happen in cosmic problems. Indeed, the full importance of the subject can be realized only when considering cosmic problems such as the earth's interior, the sun, interplanetary medium, the stars, or interstellar space. Therefore, in many ways the observations of the cosmic phenomena or large scale experiments (such as the Argus experiment) must take the place of laboratory experiments. Before we begin, we may state explicitly that the fluid considered in this study is supposed to be continuous, and that a macroscopic model will be used. By this we mean that we shall use variables similar to those in classical fluid mechanics or thermodynamics. It is known (see for instance Goldstein [24], p. 49) that this continuum theory takes no account of the net result of certain microscopic effects in an ionized gas. We neglect, for example, the relative diffusion of the electrons and ions due to the greater thermal velocities of the electrons, an effect which is independent of the presence of a magnetic field. We neglect the results of the drift of the electrons (and ions) in the direction at right angles to the electric and magnetic fields (the "Hall current"). Most important of all, we neglect the effects of the spiraling of the electrons and ions about the magnetic lines of force, and assume the collision frequency large compared with the Larmor frequency (the frequency of rotation about the magnetic lines of force). This is a serious restriction which will not be applicable at all if the density is too low (or the magnetic field too intense). At reduced density in an ionized gas, at least the scalar conductivity must be replaced by a tensor conductivity. We refer for further discussion on this material to a report by Grad [27].

As usual, charge accumulations and displacement currents are ignored. We shall require results for the fixed Newtonian frame of reference which we use for hydrodynamics. The electromagnetic variables are all supposed to be measured in emu. Clearly the equations of magnetohydrodynamics are nonlinear. One cannot, therefore, expect at the present time to do more than construct approximate solutions by appropriately simplifying assumptions.

In Part I, after writing down the basic equations of magnetohydrodynamics we present some formal apparatus of the theory *(nonlinear)*. Although the equations of magnetohydrodynamics represent a far more complex system than those of ordinary hydrodynamics, there are a number of important instances for which the two systems are mathematically identical. The existence of these exact mathematical analogies permits the immediate application of a wealth of known results in fluid dynamics to the relatively new and unexplored field of magnetohydrodynamics. Some of these analogies have already been studied by several authors including: Grad and Blank [26], Carstoiu [10], Cowling [16], Elsasser [22], Goldstein [24], and Truesdell [47]. Here, Weber's transformation is extended to the potential vector $\mathbf{A}$ and the equation for the magnetic field is integrated in terms of Lagrangian variables of hydrodynamics. These formulas are analogous to Cauchy's first integrals of the hydrodynamical equations. On the other hand, it is known that the magnetic field $\mathbf{H}$ is not a vector but a linear tensor of the second order (see for instance Weyl [52]);

here, its tensorial nature is most conspicuous. Clebsch's transformation allows a useful representation for the magnetic field. This representation leads in turn to some canonical equations for the motion of magnetic fields, of great theoretical interest (Carstoiu [13]). Part I is concluded with a theorem of decomposition of the magnetic field similar to the Cauchy-Stokes decomposition theorem in hydrodynamics.

Parts II-IV deal with oscillations of small amplitude and a linear theory is used. Part II presents the theory of hydromagnetic waves in a compressible fluid conductor. The elementary theory of hydromagnetic waves in an inviscid, incompressible, and perfectly conducting fluid embedded in a uniform magnetic field (Alfvén [1,2], Walén [51], Spitzer [44], Cowling [16], Grad [25]) indicates as a distinctive property of these waves that they do not spread out three-dimensionally around a source of disturbances, giving spherical attenuation; instead, their propagation is purely *one-dimensional* along the magnetic lines of force, and hence without attenuation. It is of interest to note here the formulations closely related to magnetohydromagnetic waves, analyzed by Poeverlein [39] for the case of electromagnetic waves in a plasma with strong magnetic field. The object of Part II is to study how this distinctive property is modified by the effect of compressibility. Compressibility effects have been considered by several authors (Herlofson [28], van de Hulst [49]; see also Sears' survey article [43]). We shall approach the problem from a different point of view, however, concentrating our attention on the vorticity field and the current density. As demonstrated here, the compressibility of a medium acts as a wave filter discriminating between components of vorticity (and current density) and passing only those directed along the (undisturbed) magnetic field. This striking property has recently been independently noticed by Grad [25], Carstoiu [9], and Lighthill [32]. Part of this theory can be found in earlier papers by Baños [5].

This mode of propagation may be a convenient source of astrophysical explanations: a disturbance at one point may originate from a disturbance at some far distant point, which has traveled between the two along a line of force. Scholte [42] related this propagation of vorticity with the giant geomagnetic pulsations, while MacDonald [36] suggested that this mode is associated with the sudden commencement of a magnetic storm. The analysis presented in Part II also seems to offer the possibility of explanation of the phenomenon of "whistlers," but there is considerably more work to be done in this direction. Rightly or wrongly, we shall not consider here any such applications which would go beyond the scope of this study. Indeed, in order to simulate and describe given *geophysical* situations one should construct a much more complex model than the one considered here. For example, one should consider a dipole magnetic field at the center of a solid sphere surrounded by a gaseous conducting atmosphere, whose density and conductivity vary with position, and study the free oscillations of the atmosphere when the whole system is rotating, with gravitation, etc., taken into account. Such a complex model is now being investigated by this writer and his group; the results obtained with applications will be presented in due time.

In Parts III and IV the dissipative effects in hydromagnetic waves are studied. The partial differential equations governing these phenomena are of mixed type, *hyperbolic-parabolic*, and this indicates that the disturbances diffuse outward as they travel along magnetic lines of force. In Part III the effect of finite conductivity alone is taken into account. This is normally the case in cosmic problems where the electrical resistance is more important than viscosity as a dissipative effect. In Part IV the combined effects of electrical conductivity and viscosity are considered. A basic account on the dissipative effects has been already published (Carstoiu [12, 14]).

The symmetrical role played in our equations by the coefficients of kinematic viscosity and magnetic viscosity is remarkable. Here one can easily see why either frictional term can be neglected depending on either $\nu|\nu_e \ll 1$ or $\nu|\nu_e \gg 1$.

The notations used in this work are generally those currently used in the pertinent literature. On rare occasions some symbols (letters) have been repeated. It is hoped that this will not produce any serious objection or misunderstanding. The bibliography at the end of this work is incomplete; the reader will find more references in each of the textbooks and articles quoted. We have referred only to those books and papers actually used in this study, or closely connected with it. We apologize for any possible omission in this ever growing domain of researches.

The theoretical investigations included in this work have been supported by contracts with the Air Force Cambridge Research Laboratories; Electronic Research Directorate,

Propagation Sciences Laboratory (Contract AF 19(604)-7487) and the National Aeronautics and Space Administration (Contract NASr-18).

It is only proper to express here my gratitude to Drs. Philip Newman and Hermann J. F. Poeverlein of the Propagation Sciences Laboratory for their constant encouragement and stimulating discussions. Dr. Henri Villat invited us to publish these researches in the "Memorial des Sciences Mathématiques," which he keeps animated; may he find here an expression of the deep gratitude of his devoted pupil and friend.

I. Magnetic Field-Fluid Dynamical Analogies

1. The Basic Equations of Magnetohydrodynamics

We shall take Cowling's "Magnetohydrodynamics" [16], Chandrasekhar's "Hydrodynamic and Hydromagnetic Stability" [15], and Goldstein's "Lectures on Fluid Mechanics" [24] as reference texts.

When the displacement currents may be neglected, Maxwell's equations are

$$\operatorname{curl} \mathbf{H} = 4\pi J \tag{1}$$

$$\operatorname{curl} \mathbf{E} = -\mu_e \frac{\partial \mathbf{H}}{\partial t} \tag{2}$$

$$\operatorname{div} \mathbf{H} = 0 \tag{3}$$

where the electromagnetic variables are measured in electromagnetic units, $\mathbf{E}$ and $\mathbf{H}$ are the intensities of the electric and magnetic fields, $\mathbf{J}$ is the current density, and μ_e is the magnetic permeability.* To complete the equations for the field, we need an equation for the current density.

Consider an electrically conducting fluid which has a conductivity σ and executes motions described by the velocity $\mathbf{v}$. The electric field it will experience is $\mathbf{E} + \mu_e \mathbf{v} \times \mathbf{H}$; thus

*The permeability is taken as unity in all practical applications (nonmagnetic materials); however, with the exception of Sections 2-8 of Part I, μ_e will be retained in our equations.

$$\mathbf{J} = \sigma(\mathbf{E} + \mu_e \mathbf{v} \times \mathbf{H}). \tag{4}$$

Equations (1)-(4) incorporate the effect of fluid motions on the electromagnetic field. The inverse effect of the field on the motions results from the ponderomotive force which the fluid elements experience by virtue of their carrying currents across magnetic lines of force. This is the Lorentz force given by

$$\mu_e \mathbf{J} \times \mathbf{H} = \frac{\mu_e}{4\pi} \operatorname{curl} \mathbf{H} \times \mathbf{H} \tag{5}$$

where Eq. (1) has been used.

Including this force among other forces acting on the fluid, we have the equation of motion

$$\rho \frac{d\mathbf{v}}{dt} = \operatorname{div} \mathbf{P} + \rho \mathbf{X} + \mu_e \mathbf{J} \times \mathbf{H} \tag{6}$$

where ρ is the density, $\mathbf{P}$ the total stress tensor, and $\mathbf{X}$ represents the external forces (per unit mass) of nonelectromagnetic origin.

In tensor notation, this equation can be written

$$\rho \left(\frac{\partial v_i}{\partial t} + v_j \frac{\partial v_i}{\partial x_j} \right) - \frac{\mu_e H_j}{4\pi} \frac{\partial H_i}{\partial x_j} = \frac{\partial P_{ij}}{\partial x_j} + \rho X_i - \frac{\partial}{\partial x_i} \left(\frac{\mu_e H^2}{8\pi} \right) \tag{7}$$

where explicitly

$$P_{ij} = -p\delta_{ij} + 2\mu e_{ij} - \frac{2}{3} \mu \delta_{ij} e_{kk}, \tag{8}$$

and where p is the isotropic pressure, μ the coefficient of viscosity and e_{ij} the rate of deformation given by

$$e_{ij} = \frac{1}{2} \left(\frac{\partial v_i}{\partial x_j} + \frac{\partial v_j}{\partial x_i} \right). \tag{9}$$

For an incompressible fluid in which μ is constant and the forces $\mathbf{X}$ derive from a potential $-\Omega$, the equation of motion simplifies to

$$\frac{\partial v_i}{\partial t} + v_j \frac{\partial v_i}{\partial x_j} - \frac{\mu_e H_j}{4\pi\rho} \frac{\partial H_i}{\partial x_j} = - \frac{\partial}{\partial x_i}\left(\Omega + \frac{p}{\rho} + \frac{\mu_e H^2}{8\pi\rho}\right) + \nu\nabla^2 v_i \qquad (10)$$

where $\nu = \mu/\rho$ denotes the kinematic viscosity.

In the general case, the equation of motion (7) has to be supplemented with the equation of continuity,

$$\frac{\partial \rho}{\partial t} + \frac{\partial}{\partial x_j}(\rho u_j) = 0, \qquad (11)$$

and the heat equation. We shall not write down the heat equation, assuming in this study that our variables do not depend on temperature.

2. The Equation of Motion for the Magnetic Field

We shall now obtain an equation of motion for the magnetic field. In view of further developments, it is convenient to introduce here the vector potential $\mathbf{A}$ and the electrostatic potential ϕ, writing in the usual way (by taking $\mu_e = 1$)

$$\mathbf{H} = \operatorname{curl} \mathbf{A} \qquad (12)$$

$$\operatorname{div} \mathbf{A} = 0 \qquad (13)$$

$$\mathbf{E} = -\frac{\partial \mathbf{A}}{\partial t} - \operatorname{grad}\phi . \qquad (14)$$

We then obtain according to Eq. (4)

$$\mathbf{J} = \sigma\left(-\frac{\partial \mathbf{A}}{\partial t} - \operatorname{grad}\phi + \mathbf{v}\times \operatorname{curl}\mathbf{A}\right). \qquad (15)$$

Substitution into Eq. (1) gives

$$\frac{\partial \mathbf{A}}{\partial t} = \mathbf{v}\times\operatorname{curl}\mathbf{A} - \operatorname{grad}\phi - \nu_e \operatorname{curl}\operatorname{curl}\mathbf{A} \qquad (16)$$

where $\nu_e = (4\pi\sigma)^{-1}$ will be designated as the magnetic viscosity (Elsasser [3]). It may be noted that ν_e like ν is of dimensions $\mathrm{cm}^2\ \mathrm{sec}^{-1}$.

Taking the curl of terms of Eq. (16) and assuming ν_e a constant, we obtain (in Cartesian coordinates)

$$\frac{\partial \mathbf{H}}{\partial t} = \operatorname{curl}(\mathbf{v} \times \mathbf{H}) + \nu_e \nabla^2 \mathbf{H} \tag{17}$$

which is the equation of motion governing magnetic field. Equation (17) is general; it is not restricted either to incompressible fluids or to inviscid fluids. There is a complete mathematical analogy between Eq. (17) and the equation governing the vorticity in a viscous incompressible fluid. This analogy will be studied in greater detail in subsequent analyses.

The case in which the electrical conductivity of the medium may be considered as infinite is of particular interest in cosmic electrodynamics. The magnetic viscosity is then zero, and Eqs. (16) and (17) reduce, respectively, to

$$\frac{\partial \mathbf{A}}{\partial t} = \mathbf{v} \times \operatorname{curl} \mathbf{A} - \operatorname{grad} \phi \tag{18}$$

and

$$\frac{\partial \mathbf{H}}{\partial t} = \operatorname{curl}(\mathbf{v} \times \mathbf{H}). \tag{19}$$

Equations (18) and (19), particularly the latter, have been the object of considerable research in the literature. We may notice its full analogy with the Helmholtz equation for the vorticity. This immediately permits the application of, *mutatis mutandis*, the classical and elegant results of the theory of vorticity to the magnetic field. In particular, it follows at once that the lines of magnetic force move with the fluid. For further details of this analogy we refer to Goldstein's book [24] (see pp. 76-77).

3. The Elsasser-Carstoiu Theorem

Equation (19) may be put in a different form which is reminiscent of the Cauchy-Stokes decomposition of an arbitrary instantaneous continuous motion of a fluid. Equivalent to the basic equation (19) is

$$\frac{d}{dt}\left(\frac{H_i}{\rho}\right) = \frac{H_j}{\rho}\frac{\partial v_i}{\partial x_j} \tag{20}$$

where the equation of continuity (11) has been used. This can be rewritten

$$\frac{d}{dt}\left(\frac{H_i}{\rho}\right) = \frac{1}{2}\frac{H_j}{\rho}\left(\frac{\partial v_i}{\partial x_j} - \frac{\partial v_j}{\partial x_i}\right) + \frac{1}{2}\frac{H_j}{\rho}\left(\frac{\partial v_i}{\partial x_j} + \frac{\partial v_j}{\partial x_i}\right) \tag{21}$$

where, besides the rate of deformation e_{ij}, the vorticity

$$\omega_{ij} = \frac{1}{2}\left(\frac{\partial v_i}{\partial x_j} - \frac{\partial v_j}{\partial x_i}\right) \tag{22}$$

appears under its tensor components. We can write

$$\frac{d}{dt}\left(\frac{H_i}{\rho}\right) = \frac{H_j}{\rho}\omega_{ij} + \frac{1}{2\rho}\frac{\partial G}{\partial H_i} \tag{23}$$

or, in vector notation,

$$\frac{d}{dt}\left(\frac{\mathbf{H}}{\rho}\right) = \boldsymbol{\omega} \times \frac{\mathbf{H}}{\rho} + \frac{1}{2\rho}\,\mathrm{Grad}_{H_i} G \tag{24}$$

where we set

$$G = e_{ij}H_iH_j, \tag{25}$$

and Grad_{H_i} is taken with respect to H_i (x_i fixed).

Equation (24) shows that the rate of change of the magnetic field may be conceived of as being made up of two parts. The first part expresses a rotation of the field with the fluid particle; and the second part shows that the terminus of $\mathbf{H}$ is moving in the direction of the normal to that quadric of the system

$$e_{ij}X_iX_j = \text{constant} \tag{26}$$

on which its terminus lies.

In this form the theorem has been stated by this writer [10]. An integral formulation closely related to this was first given by Elsasser [19, 20].

Equation (23) is to be contrasted with the form in which can be put the vorticity equation (Carstoiu [8]), namely,

$$\frac{d}{dt}\left(\frac{\omega_i}{\rho}\right) = \frac{1}{2\rho}\frac{\partial F}{\partial \omega_i} \tag{27}$$

where

$$F = e_{ij}\omega_i\omega_j . \tag{28}$$

It takes a similar form only in two cases: (a) $\boldsymbol{\omega}$ parallel to $\mathbf{H}$; (b) irrotational motion.

A remarkable formal property of the magnetic field can readily be inferred from Eq. (24). We have

$$\frac{1}{2}\frac{d}{dt}\left(\frac{H^2}{\rho^2}\right) = \frac{G}{\rho^2} . \tag{29}$$

Hence, H/ρ will be a constant with respect to time for any material particle if and only if the magnetic field lies on the asymptotic cone of the quadric of deformation at the point x_i. This constitutes the analog for the magnetic field of a vorticity property as has been pointed out by the writer [8].

4. Rigid Rotation of Material

As a limiting case assume a rigid rotation of the material. Since a rigid rotation is characterized by $e_{ij} \equiv 0$, Eq. (24) reduces to

$$\frac{d\mathbf{H}}{dt} = \boldsymbol{\Omega}(t) \times \mathbf{H} \tag{30}$$

where $\boldsymbol{\Omega}$ is a function of time only. Expanding the left-hand side of Eq. (30), this becomes

$$\frac{\partial \mathbf{H}}{\partial t} + [(\boldsymbol{\Omega}\times \mathbf{r}) \cdot \nabla]\mathbf{H} = \boldsymbol{\Omega}\times \mathbf{H}. \tag{31}$$

Now

$$[(\boldsymbol{\Omega}\times \mathbf{r}) \cdot \nabla]\mathbf{H} = \operatorname{grad}[(\boldsymbol{\Omega}\times \mathbf{r}) \cdot \mathbf{H}] + \boldsymbol{\Omega}\times \mathbf{H} - 4\pi(\boldsymbol{\Omega}\times \mathbf{r}) \times \mathbf{J}. \tag{32}$$

Equation (31) can consequently be rewritten as

$$\frac{\partial \mathbf{H}}{\partial t} + \operatorname{grad}[(\boldsymbol{\Omega} \times \mathbf{r}) \cdot \mathbf{H}] - 4\pi(\boldsymbol{\Omega} \times \mathbf{r}) \times \mathbf{J} = 0. \tag{33}$$

The curl of terms of this equation gives

$$\frac{\partial \mathbf{J}}{\partial t} + [(\boldsymbol{\Omega} \times \mathbf{r}) \cdot \nabla]\mathbf{J} - \boldsymbol{\Omega} \times \mathbf{J} = 0; \tag{34}$$

that is,

$$\frac{d\mathbf{J}}{dt} = \boldsymbol{\Omega} \times \mathbf{J}. \tag{35}$$

Equations (30) and (35) show at once that

$$H = \text{constant} \tag{36}$$

$$J = \text{constant} \tag{37}$$

for any material point. On the other hand, scalar multiplication of terms of Eq. (30) by $\mathbf{J}$ yields

$$\mathbf{J} \cdot \frac{d\mathbf{H}}{dt} = -\mathbf{H} \cdot (\boldsymbol{\Omega} \times \mathbf{J}) = -\mathbf{H} \cdot \frac{d\mathbf{J}}{dt}; \tag{38}$$

that is,

$$\mathbf{H} \cdot \mathbf{J} = \text{constant}. \tag{39}$$

The latter shows, by virtue of Eqs. (36) and (37), that the angle between $\mathbf{H}$ and $\mathbf{J}$ is constant. In particular, if $\mathbf{H}$ and $\mathbf{J}$ are parallel at $t = 0$ for any material point, then they are always parallel for that point, i.e., *an initially force-free field is preserved* during the motion. This property may also be obtained as follows. We have

$$\mathbf{J} \times \frac{d\mathbf{H}}{dt} = \mathbf{J} \times (\boldsymbol{\Omega} \times \mathbf{H}) = (\mathbf{J} \cdot \mathbf{H})\boldsymbol{\Omega} - (\mathbf{J} \cdot \boldsymbol{\Omega})\mathbf{H} \tag{40}$$

and

$$\mathbf{H} \times \frac{d\mathbf{J}}{dt} = \mathbf{H} \times (\boldsymbol{\Omega} \times \mathbf{J}) = (\mathbf{J} \cdot \mathbf{H})\boldsymbol{\Omega} - (\mathbf{H} \cdot \boldsymbol{\Omega})\mathbf{J}. \tag{41}$$

Hence

$$\frac{d}{dt}(\mathbf{J}\times\mathbf{H}) = (\mathbf{H}\cdot\Omega)\mathbf{J} - (\mathbf{J}\cdot\Omega)\mathbf{H} = \Omega\times(\mathbf{J}\times\mathbf{H}) \tag{42}$$

which shows that $\mathbf{J}\times\mathbf{H}$ is carried with the material; if $\mathbf{J}\times\mathbf{H}$ vanishes at $t = 0$, then it always remains zero.

Assuming the existence of a force-free field (the other body forces having a potential), this implies that the angular velocity vector is a constant; for by taking the curl of terms of Eq. (6), this gives

$$\frac{d\mathbf{\Omega}}{dt} = \frac{\partial\mathbf{\Omega}}{\partial t} = (\mathbf{\Omega}\cdot\nabla)(\mathbf{\Omega}\times\mathbf{r}) = \mathbf{\Omega}\times\mathbf{\Omega} = 0 \tag{43}$$

which asserts that Ω = constant. A brief account of these results has been given earlier by the writer [10].

5. The Analog of the Clebsch Transformation

Another matter of interest in cosmic electrodynamics is the analog of the Clebsch transformation of the hydrodynamical equations (see for instance, Lamb [30], p. 248). Putting

$$\mathbf{A} = \operatorname{grad} S + \Phi \operatorname{grad}\Psi, \tag{44}$$

one has

$$\mathbf{H} = \operatorname{curl}(\Phi\operatorname{grad}\Psi) = \operatorname{grad}\Phi\times\operatorname{grad}\Psi. \tag{45}$$

The representation (45) is identical to that given by Sweet [46] (see also Dungey [18]), with the exception of a factor F, function of Φ and Ψ only, which appears on the right-hand side of (45) in Sweet's representation. It can, however, be shown (see Lamb [30]) that these two representations are equivalent.

The immediate consequences of representation (45), namely,

$$\mathbf{H}\cdot\operatorname{grad}\Phi = \mathbf{H}\cdot\operatorname{grad}\Psi = 0, \tag{46}$$

show that the magnetic field is tangent to the surfaces Φ = constant and Ψ = constant, which we shall call *surfaces of force* (a surface composed wholly of lines of force), and which correspond to vortex surfaces in hydrodynamics. It is evident that their intersections are the lines of force.

6. Hamilton's Form of the Equation of Motion for the Magnetic Field

Let us now come back to Eq. (18) and substitute therein the value of $\mathbf{A}$ given by (44). We have

$$\frac{\partial}{\partial t}(\mathrm{grad}\, S + \Phi\, \mathrm{grad}\, \Psi) = \mathbf{v} \times (\mathrm{grad}\, \Phi \times \mathrm{grad}\, \Psi) - \mathrm{grad}\, \phi$$

$$= (\mathbf{v} \cdot \mathrm{grad}\, \Psi)\, \mathrm{grad}\, \Phi$$

$$- (\mathbf{v} \cdot \mathrm{grad}\, \Phi)\, \mathrm{grad}\, \Psi - \mathrm{grad}\, \phi. \tag{47}$$

which can be written

$$\frac{d\Phi}{dt}\, \mathrm{grad}\, \Psi - \frac{d\Psi}{dt}\, \mathrm{grad}\, \Phi = -\, \mathrm{grad}\, \mathcal{H} \tag{48}$$

where

$$\mathcal{H} = \frac{\partial S}{\partial t} + \Phi \frac{\partial \Psi}{\partial t} + \phi. \tag{49}$$

Scalar multiplication of terms of Eq. (48) by $\mathrm{grad}\, \phi \times \mathrm{grad}\, \Psi$ gives

$$\mathrm{grad}\, \mathcal{H} \cdot (\mathrm{grad}\, \Phi \times \mathrm{grad}\, \Psi) = 0\,; \tag{50}$$

that is, the Jacobian

$$\frac{\partial(\mathcal{H}, \Phi, \Psi)}{\partial(x, y, z)} = 0. \tag{51}$$

This shows that $\mathcal{H}$ is the form $\mathcal{H}(\Phi, \Psi, t)$. Hence

$$\mathrm{grad}\, \mathcal{H} = \frac{\partial \mathcal{H}}{\partial \Phi}\, \mathrm{grad}\, \Phi + \frac{\partial \mathcal{H}}{\partial \Psi}\, \mathrm{grad}\, \Psi\,. \tag{52}$$

Comparison of Eqs. (48) and (52) gives at once the Hamiltonian system

$$\frac{d\Phi}{dt} = -\frac{\partial\mathcal{H}}{\partial\Psi}, \qquad \frac{d\Psi}{dt} = \frac{\partial\mathcal{H}}{\partial\Phi}. \tag{53}$$

Equations (53) are analogous to Stuart's equations in hydrodynamics (see Lamb [30]).

7. The Analog of the Weber Transformation

Equation (18) may be written

$$\frac{d\mathbf{A}}{dt} = -\mathbf{A} \times \operatorname{curl} \mathbf{v} - (\mathbf{A} \cdot \nabla)\mathbf{v} + \operatorname{grad}(\mathbf{A} \cdot \mathbf{v} - \phi) \tag{54}$$

or, in tensor motation,

$$\frac{dA_i}{dt} + A_j \frac{\partial v_j}{\partial x_i} = \frac{\partial}{\partial x_i}(A_j v_j - \phi). \tag{55}$$

We now introduce the Lagrangian variables, i.e., label a material particle by the parameters x_α^0 which indicate its position at the time $t = 0$. We designate by A_α^0, v_α^0, etc., initial conditions, i.e., the values taken by these variables at $t = 0$. Multiplying the terms of Eq. (55) by $\partial x_i/\partial x_\alpha^0$, we have

$$\frac{\partial A_i}{\partial t} \frac{\partial x_i}{\partial x_\alpha^0} + A_i \frac{\partial}{\partial t} \frac{\partial x_i}{\partial x_\alpha^0} = \frac{\partial}{\partial x_\alpha^0}(A_i v_i - \phi), \tag{56}$$

so that

$$\frac{\partial}{\partial t}\left(A_i \frac{\partial x_i}{\partial x_\alpha^0}\right) = \frac{\partial}{\partial x_\alpha^0}(A_i v_i - \phi). \tag{57}$$

Integration of Eq. (57) with respect to time between the limits 0 and t yields

$$A_i \frac{\partial x_i}{\partial x_\alpha^0} - A_\alpha^0 = \frac{\partial \chi}{\partial x_\alpha^0}, \tag{58}$$

if we write

$$\chi = \int_0^t (A_i v_i - \phi)\, dt . \tag{59}$$

Equation (58) gives a mathematical analog of the Weber's transformation (see Lamb [30], p. 14). We recall that the latter implies an inviscid fluid in which p is a function of ρ only, with body forces conservative. Equation (58) is quite general. In particular, it is not restricted to inviscid fluids. The dependent variables have different signification in magnetohydromagnetic and fluid mechanics. We have, however, in both domains the same functional relationship

$$A_i\, dx_i - \overset{0}{A}_a\, d\overset{0}{x}_a = d\chi . \tag{60}$$

Hence, we have the following conservation theorems analogous to Kelvin and Helmholtz theorems. First

$$\oint A_i\, dx_i = \oint A_a^0\, dx_a^0 \tag{61}$$

which is the analog of Kelvin's theorem and which states that the circulation of vector $\mathbf{A}$ in any circuit moving with the fluid is constant for all time.

In the second place, by Stokes' theorem

$$\int H_n\, dS = \int H_n^0\, dS^0 \tag{62}$$

which is the counterpart of the Helmholtz-Kelvin theorem and which states that the integral of the normal component of $\mathbf{H}$ over a surface S bounded by a closed curve remains constant as we follow the surface S with the motion of the fluid elements constituting it.

Formulas (61) and (62) show in the simplest manner that a *surface of force* moves with the fluid. For, consider at $t = 0$ a surface of force S^0. For such a surface $H_n^0 = 0$. The circulation of vector $\overset{0}{\mathbf{A}}$ round any circuit C^0 is zero by Stokes's theorem at $t = 0$. Now let S^0 and C^0 move with the fluid. According to (61) the circulation round C remains zero, and this is true for any circuit. Hence, $\int H_n\, dS = 0$ for all portions of S, and $H_n = 0$ at every point of S at all subsequent times. Hence S remains a surface of force. An alternate proof of this theorem will be given in the subsequent analysis.

8. The Analog of the Cauchy Equations and Conservation Theorems

Cauchy's equations are found in the classical books on fluid mechanics. Their counterparts for the magnetic field have been investigated especially by Elsasser [22] and Goldstein [24].

Now, it is known that **H** *(like ω) is not a vector, but a linear tensor of the second order* (see for instance Weyl [52]) defined by

$$H_{ij} = \frac{\partial A_i}{\partial x_j} - \frac{\partial A_j}{\partial x_i} . \tag{63}$$

Its true tensorial character will appear clearly in this analysis. Let us return to the basic equation (58) and differentiate with respect to $\overset{0}{x}_\beta$; we have

$$\frac{\partial A_i}{\partial x_j} \frac{\partial x_i}{\partial x_\alpha^0} \frac{\partial x_j}{\partial x_\beta^0} + A_i \frac{\partial^2 x_i}{\partial x_\alpha^0 \partial x_\beta^0} - \frac{\partial A_\alpha^0}{\partial x_\beta^0} = \frac{\partial^2 \chi}{\partial x_\alpha^0 \partial x_\beta^0} . \tag{64}$$

By simple changes of free and dummy indices in the above equation, and subtraction of the two results, we have, in consequence of (63),

$$H_{ij} \frac{\partial x_i}{\partial x_\alpha^0} \frac{\partial x_j}{\partial x_\beta^0} - H_{\alpha\beta}^0 = 0. \tag{65}$$

Equation (65) is of fundamental importance. It may be interpreted in two ways: (a) it gives the whole history of H_{ij} in terms of its initial value; (b) we can regard (65) as giving, at any time, the components of H_{ij} in the two systems of coordinates x_i and x_α^0 where

$$x_i = x_i \left(x_1^0, x_2^0, x_3^0, t \right) , \tag{66}$$

and t is a parameter. This equation shows plainly that H_{ij} is tensor of second order. Equation (65) constitutes a radical departure from the conventional equations for the magnetic field found in the textbooks and is valid in any number of dimensions. Its counterpart for vorticity can be found in an earlier work of the writer [8]. In rectangular Cartesian coordinates it takes the form

$$\frac{1}{2} H_{ij} \frac{\partial(x_i, x_j)}{\partial(x_\alpha^0, x_\beta^0)} - H_{\alpha\beta}^0 = 0. \tag{67}$$

If we put $i, j; \alpha, \beta = 1, 2, 3$; and $H_{32} = H_1$, $H_{13} = H_2$, $H_{21} = H_3$, etc., we obtain

$$H_1 \frac{\partial(x_3, x_2)}{\partial(x_2^0, \partial x_3^0)} + H_2 \frac{\partial(x_1, x_3)}{\partial(x_2^0, x_3^0)} + H_3 \frac{\partial(x_2, x_1)}{\partial(x_2^0, x_3^0)} = H_1^0, \quad \text{etc.} \tag{68}$$

Solving for H_1, H_2, H_3 and using the Lagrangian equation of continuity, namely,

$$\rho \frac{\partial(x_1, x_2, x_3)}{\partial(x_1^0, x_2^0, x_3^0)} = \rho^0, \tag{69}$$

we obtain

$$\frac{H_1}{\rho} = \frac{H_1^0}{\rho^0} \frac{\partial x_1}{\partial x_1^0} + \frac{H_2^0}{\rho^0} \frac{\partial x_1}{\partial x_2^0} + \frac{H_3^0}{\rho^0} \frac{\partial x_1}{\partial x_3^0}, \quad \text{etc.} \tag{70}$$

These are the counterparts for the magnetic field of Cauchy's equations.

Several important consequences for the field follow from Eq. (65). We shall now consider some of them.

(a) Flux of the Magnetic Field

As is well known, a material element of surface dS_{ij} is related to its initial value $dS_{\alpha}^0 s$ by the equation

$$dS_{ij} = dS^0_{\alpha\beta} \frac{\partial x_i}{\partial x^0_\alpha} \frac{\partial x_j}{\partial x^0_\beta} . \tag{71}$$

Multiplying both sides of (65) by $dS^0_{\alpha\beta}$ we obtain

$$H_{ij}\, dS_{ij} = H^0_{\alpha\beta}\, dS^0_{\alpha\beta} . \tag{72}$$

Hence

$$\int H_{ij}\, dS_{ij} = \int H^0_{\alpha\beta}\, dS^0_{\alpha\beta} \tag{73}$$

and

$$\oint A_i\, dx_i = \oint A^0_\alpha\, dx^0_\alpha . \tag{74}$$

Vice versa, (65) may be obtained by starting with (73) and using formula (71). This gives

$$\int \left(H_{ij} \frac{\partial x_i}{\partial x^0_\alpha} \frac{\partial x_j}{\partial x^0_\beta} - H^0_{\alpha\beta} \right) dS^0_{\alpha\beta} = 0. \tag{75}$$

The result (65) now becomes apparent by observing that $dS^0_{\alpha\beta}$ may be selected arbitrarily (see also Goldstein [24], pp. 74-75).

(b) *Surfaces of Force*

A material surface $x_i = x_i(\lambda, \mu)$ is defined by any initial surface $x^0_\alpha = x^0_\alpha(\lambda, \mu)$ and by the equations $x_i = x_i\left[x^0_\alpha(\lambda, \mu), t\right]$. At all times a given material point on this surface corresponds to the parameters λ, μ.

Consider again Eq. (65) and multiply both sides of this equation by $(\partial x^0_\alpha/\partial\lambda)(\partial x^0_\beta/\partial\mu)$. We have

$$H_{ij} \frac{\partial x_i}{\partial x^0_\alpha} \frac{\partial x_j}{\partial x^0_\beta} \frac{\partial x^0_\alpha}{\partial\lambda} \frac{\partial x^0_\beta}{\partial\mu} = H^0_{\alpha\beta} \frac{\partial x^0_\alpha}{\partial\lambda} \frac{\partial x^0_\beta}{\partial\mu} ; \tag{76}$$

that is,

$$H_{ij}\frac{\partial x_i}{\partial\lambda}\frac{\partial x_j}{\partial\mu} = H^0_{\alpha\beta}\frac{\partial x^0_\alpha}{\partial\lambda}\frac{\partial x^0_\beta}{\partial\mu}\,. \tag{77}$$

It is obvious that the vanishing of either side of (77) implies the vanishing of the other. But the vanishing of either side expresses geometrically that **H** is contained in the tangent plane at the respective surface. Hence, if our initial surface is a surface of force, it will continue to be so by virtue of (77).

This geometrical property has an interesting analytical counterpart which may be related to Clebsch's transformation (see Lamb [30] and Appell [3]).

Consider the elementary flux

$$A_i\,dx_i \tag{78}$$

and try to find if there are surfaces on which (78) becomes an exact differential (in two variables). Now, on any surface one can express x_i as functions of two variables λ and μ and one has

$$dx_i = \frac{\partial x_i}{\partial\lambda}\,d\lambda + \frac{\partial x_i}{\partial\mu}\,d\mu. \tag{79}$$

Hence

$$A_i\,dx_i = A_i\frac{\partial x_i}{\partial\lambda}\,d\lambda + A_i\frac{\partial x_i}{\partial\mu}\,d\mu. \tag{80}$$

This is an exact differential if and only if

$$\frac{\partial}{\partial\mu}\left(A_i\frac{\partial x_i}{\partial\lambda}\right) - \frac{\partial}{\partial\lambda}\left(A_i\frac{\partial x_i}{\partial\mu}\right) = 0 \tag{81}$$

which gives

$$\frac{\partial A_i}{\partial\mu}\frac{\partial x_i}{\partial\lambda} - \frac{\partial A_j}{\partial\lambda}\frac{\partial x_j}{\partial\mu} = 0. \tag{82}$$

Now

$$\frac{\partial A_i}{\partial \mu} = \frac{\partial A_i}{\partial x_j} \frac{\partial x_j}{\partial \mu} \tag{83}$$

and

$$\frac{\partial A_j}{\partial \lambda} = \frac{\partial A_j}{\partial x_i} \frac{\partial x_i}{\partial \lambda} . \tag{84}$$

Substitution of (83) and (84) into (82) gives

$$\left(\frac{\partial A_i}{\partial x_j} - \frac{\partial A_j}{\partial x_i}\right) \frac{\partial x_i}{\partial \lambda} \frac{\partial x_j}{\partial \mu} = 0; \tag{85}$$

that is,

$$H_{ij} \frac{\partial x_i}{\partial \lambda} \frac{\partial x_j}{\partial \mu} = 0 \tag{86}$$

which shows that the surfaces in question are surfaces of force.

9. Nonconservation of the Current Density

It is of interest to notice that the current density **J** is not preserved during the motion. We may write

$$4\pi J_i = \frac{\partial H_{ki}}{\partial x_k} = \frac{\partial H_{1i}}{\partial x_1} + \frac{\partial H_{2i}}{\partial x_2} + \frac{\partial H_{3i}}{\partial x_3} , \tag{87}$$

for one has, for instance,

$$4\pi J_1 = \frac{\partial H_{21}}{\partial x_1} + \frac{\partial H_{31}}{\partial x_3} = \frac{\partial H_3}{\partial x_2} - \frac{\partial H_2}{\partial x_3} , \quad \text{etc.}, \tag{88}$$

which give precisely Eq. (1). At the initial instant, we have

$$4\pi J_\alpha^0 = - \frac{\partial H^0_{\alpha\beta}}{\partial x^0_\beta} . \tag{89}$$

Now let us come back to the basic equation (65) and differentiate with respect to x^0_β. This gives

$$\frac{\partial H_{ij}}{\partial x_k} \frac{\partial x_i}{\partial x^0_\alpha} \frac{\partial x_j}{\partial x^0_\beta} \frac{\partial x_k}{\partial x^0_\beta} + H_{ij} \frac{\partial}{\partial x^0_\beta}\left(\frac{\partial x_i}{\partial x^0_\alpha} \frac{\partial x_j}{\partial x^0_\beta}\right) + h\pi J^0_\alpha = 0. \tag{90}$$

Thus, an initial current density J^0_α attached to a fluid particle is not preserved with this particle (as is the magnetic field). Whether or not there exist physical quantitites other than the magnetic field which may be preserved with the fluid (such as the magnetic field does) is an open question.

10. Changes in $w_i = H_{ij}\, dx_j$ Induced by the Motion

We shall study in this section the infinitesimal vector

$$w_i = H_{ij}\, dx_j \tag{91}$$

associated at time t with a given infinitesimal field dx_i which changes in time according to the rule

$$dx_i = \frac{\partial x_i}{\partial x^0_\alpha}\, dx^0_\alpha. \tag{92}$$

Thus, if the field dx_i is at one instant tangent to congruence of material curves, it will remain so for all time. Setting $w_i = 0$ yields a differential system for the lines of force.

Multiplying (65) by dx^0_β yields

$$w_i \frac{\partial x_i}{\partial x^0_\alpha} - w^0_\alpha = 0. \tag{93}$$

This equation shows that the vector w_i moves with the fluid or is "frozen" into the material medium. In particular, if w^0_α vanishes, then w_i is always zero; i.e., a material line, initially a line of force, always remains a line of force.

Equivalent to (93) is

$$w_i = w^0_\alpha \frac{\partial x^0_\alpha}{\partial x_i}. \tag{94}$$

The latter yields three equations of the type

$$\frac{w_1}{\rho} = \frac{w_1^0}{\rho^0} \frac{\partial(x_2, x_3)}{\partial(x_2^0, \partial x_3^0)} + \frac{w_2^0}{\rho^0} \frac{\partial(x_2, x_3)}{\partial(x_3^0, x_1^0)} + \frac{w_3^0}{\rho^0} \frac{\partial(x_2, x_3)}{\partial(x_1^0, x_2^0)} \tag{95}$$

which may be compared with the basic equations (68). Also, a differential system can be derived from (93) by applying to the latter the operator d/dt. We have

$$\frac{dw_i}{dt} \frac{\partial x_i}{\partial x_\alpha^0} + w_i \frac{\partial v_i}{\partial x_\alpha^0} = 0. \tag{96}$$

Multiplying by $\partial x_\alpha^0 / \partial x_k$ we obtain

$$\frac{dw_k}{dt} + w_i \frac{\partial v_i}{\partial x_k} = 0 \tag{97}$$

which may be compared with the basic equation (20).

From Eqs. (95), it follows that a material surface initially normal to w_α^0 always remains normal to w_i. For setting

$$x_\alpha^0 = x_\alpha^0(\lambda, \mu), \tag{98}$$

we have initially

$$\frac{w_1^0}{\dfrac{\partial(x_2^0, x_3^0)}{\partial(\lambda, \mu)}} = \frac{w_2^0}{\dfrac{\partial(x_3^0, x_1^0)}{\partial(\lambda, \mu)}} = \frac{w_3^0}{\dfrac{\partial(x_1^0, x_2^0)}{\partial(\lambda, \mu)}} = \frac{w_0}{\sqrt{E^0 G^0 - (F^0)^2}} \tag{99}$$

where E^0, F^0, G^0 are the Gaussian coefficients for this surface. Substitution of w_α^0 into Eqs. (95) yields

$$\frac{w_1}{\dfrac{\partial(x_2, x_3)}{\partial(\lambda, \mu)}} = \frac{w_2}{\dfrac{\partial(x_3, x_1)}{\partial(\lambda, \mu)}} = \frac{w_3}{\dfrac{\partial(x_1, x_2)}{\partial(\lambda, \mu)}} = \frac{\rho}{\rho^0} \frac{w^0}{\sqrt{E^0 G^0 - (F^0)^2}}$$

$$= \frac{w}{\sqrt{EG - F^2}}, \tag{100}$$

which concludes the proof. It also follows that

$$\frac{\rho}{\rho_0}\,\frac{dS}{dS^0} = \frac{w}{w^0} \tag{101}$$

where dS is a material element of area.

11. A New Pseudo-Tensor, Γ_{ij}, Derived from the Magnetic Field

We now use vector notation for the magnetic field H_i, and introduce the quantity

$$\Gamma_{ij} = \frac{1}{2}\left(\frac{\partial H_i}{\partial x_j} + \frac{\partial H_j}{\partial x_i}\right). \tag{102}$$

This is obviously mathematically derived from the magnetic field in the same way as the rate of deformation e_{ij} [formula (9)] is derived from material velocity. Its physical significance is not obvious (it has dimensions of current density) but following the Cauchy-Stokes decomposition theorem in hydrodynamics we can say this: if one compares, at the time t, the magnetic fields at two infinitely near material particles at x_i and $x_i' = x_i + \delta x_i$, then one has

$$H_i' = H_i + \frac{\partial H_i}{\partial x_j}\,\delta x_j \tag{103}$$

which can be written as

$$H_i' = H_i + \frac{1}{2}\left(\frac{\partial H_i}{\partial x_j} - \frac{\partial H_j}{\partial x_i}\right)\delta x_j + \frac{1}{2}\left(\frac{\partial H_i}{\partial x_j} + \frac{\partial H_j}{\partial x_i}\right)\delta x_j \tag{104}$$

or, in vector notation,

$$\mathbf{H}' = \mathbf{H} + 2\pi\mathbf{J}\times\delta\mathbf{r} + \frac{1}{2}\,\mathrm{Grad}_{\delta x_i}\mathfrak{J} \tag{105}$$

where we set

$$\mathfrak{J} = \Gamma_{ij}\,\delta x_i\,\delta x_j, \tag{106}$$

and the gradient is taken with respect to δx_i (x_i being fixed). Now, while the second part on the right-hand side of Eq. (105) expresses Laplace's law plainly, the last part is somewhat more difficult to interpret and requires further investigation. Assuming

$$\Gamma_{ij} = 0, \tag{107}$$

this gives

$$\mathbf{H} = \mathbf{H}_0(t) + 2\pi\mathbf{J}(t) \times \mathbf{r}. \tag{108}$$

It can easily be verified that formula (108) implies

$$\operatorname{curl} \mathbf{H} = 4\pi\mathbf{J} \tag{109}$$

$$\operatorname{div} \mathbf{H} = 0. \tag{110}$$

The fields satisfying (108) constitute singular fields, and should be more closely investigated.

II. Magnetohydrodynamic Waves in a Compressible Fluid

12. Preliminaries—The Case of an Incompressible Fluid

We begin with a discussion of magnetohydrodynamic waves in the case of an incompressible fluid. Alfvén [1, 2] has shown that these waves can travel along magnetic lines of force in a conducting material. A rigorous theory of magnetohydrodynamic waves, however, was first considered by Walén [51]. Here, Walén's results will be given briefly. Consider an infinite mass of uniform fluid at rest embedded in a uniform magnetic field. We assume this fluid to be an inviscid, incompressible, and perfectly conducting material. The latter assumption can be justified for a sufficiently large-scale disturbance. Assume that as a result of a perturbation, a velocity field $\mathbf{v}$ is produced in a certain region, and that the magnetic field becomes $\mathbf{H}_0 + \mathbf{h}$. The equations giving the variations in $\mathbf{v}$ and $\mathbf{h}$ are

$$\rho_0 \frac{\partial \mathbf{v}}{\partial t} = -\operatorname{grad} p + \rho_0 g + \frac{\mu_e}{4\pi} \operatorname{curl} \mathbf{h} \times (\mathbf{H}_0 + \mathbf{h}) \tag{1}$$

$$\frac{\partial \mathbf{h}}{\partial t} = \operatorname{curl} \mathbf{v} \times (\mathbf{H}_0 + \mathbf{h}) \tag{2}$$

where we have included the gravitational potential

$$\mathbf{g} = -\operatorname{grad} \psi \tag{3}$$

and ρ_0 is the uniform density of our fluid, and the term $(\mathbf{v} \cdot \nabla)\mathbf{v}$ has been omitted.

Now, since $\mathbf{H}_0$ = constant, and

$$\operatorname{div} \mathbf{v} = 0 \tag{4}$$

$$\operatorname{div} \mathbf{h} = 0, \tag{5}$$

the equations simplify to

$$\rho_0 \frac{\partial \mathbf{v}}{\partial t} = -\operatorname{grad} \left(p + \frac{\mu_e H_0 \cdot \mathbf{h}}{4\pi} + \rho_0 \psi \right) + \frac{\mu_0}{4\pi} (\mathbf{H}_0 \cdot \nabla)\mathbf{h} \tag{6}$$

$$\frac{\partial \mathbf{h}}{\partial t} = (\mathbf{H}_0 \cdot \nabla)\mathbf{v} \tag{7}$$

by neglecting squares and products of the small quantities $\mathbf{h}, \mathbf{v}$. Take the divergence of Eq. (6); we have

$$\operatorname{div} \operatorname{grad} \left(p + \frac{\mu_e \mathbf{H}_0 \cdot \mathbf{h}}{4\pi} + \rho_0 \psi \right) = 0. \tag{8}$$

In Cartesian coordinates Eq. (8) reduces to

$$\nabla^2 \left(p + \frac{\mu_e \mathbf{H}_0 \cdot \mathbf{h}}{4\pi} + \rho_0 \psi \right) = 0. \tag{9}$$

Since $p + (\mu_e \mathbf{H}_0 \cdot \mathbf{h}/4\pi) + \rho_0 \psi$ has no singularities and is bounded,

$$p + \frac{\mu_e \mathbf{H}_0 \cdot \mathbf{h}}{4\pi} + \rho_0 \psi = \text{constant}. \tag{10}$$

Hence, Eq. (6) becomes

$$4\pi\rho_0 \frac{\partial \mathbf{v}}{\partial t} = \mu_e(\mathbf{H}_0 \cdot \nabla)\mathbf{h}. \tag{11}$$

For simplicity take Oz parallel to $\mathbf{H}_0$. Then Eqs. (11) and (7) become

$$4\pi\rho_0 \frac{\partial \mathbf{v}}{\partial t} = \mu_e H_0 \frac{\partial \mathbf{h}}{\partial z} \tag{12}$$

$$\frac{\partial \mathbf{h}}{\partial t} = H_0 \frac{\partial \mathbf{v}}{\partial z} \tag{13}$$

Hence, by cross-differentiation,

$$\frac{\partial^2 \mathbf{v}}{\partial t^2} = A_0^2 \frac{\partial^2 \mathbf{v}}{\partial z^2} \tag{14}$$

$$\frac{\partial^2 \mathbf{h}}{\partial t^2} = A_0^2 \frac{\partial^2 \mathbf{h}}{\partial z^2} \tag{15}$$

where

$$A_0^2 = \frac{\mu_e H_0^2}{4\pi\rho_0} \tag{16}$$

is the Alfvén phase velocity, named so in honor of its discoverer. Thus the disturbance can be expressed as the sum of two sets of waves traveling with velocities $\pm A$, in the z-direction; i.e., along the lines of force of the undisturbed field. These waves are called magnetohydrodynamic (m.h.) waves, or hydromagnetic waves.

After the two waves have separated we have, in either of the waves,

$$\frac{\partial \mathbf{v}}{\partial t} = \pm A_0 \frac{\partial \mathbf{v}}{\partial z}, \tag{17}$$

the sign depending on the direction of propagation of the wave considered. Comparison of Eqs. (13) and (17) gives

$$\mathbf{h} = \pm \frac{H_0 \mathbf{v}}{A_0} = \pm \sqrt{\frac{4\pi\rho_0}{\mu_e}}\, \mathbf{v}. \tag{18}$$

Before going further, we note that in considering the propagation described by Alfvén, the velocity $\mathbf{v}$ and the magnetic field $\mathbf{h}$ can be replaced by the vorticity ω and the current density $\mathbf{j}$, respectively $[\mathbf{j} = (1/4\pi)\operatorname{curl}\mathbf{h}]$, for one has similar equations for these quantities, namely,

$$\frac{\partial^2 \omega}{\partial t^2} = A_0^2 \frac{\partial^2 \omega}{\partial z^2} \tag{19}$$

$$\frac{\partial^2 \mathbf{j}}{\partial t^2} = A_0^2 \frac{\partial^2 \mathbf{j}}{\partial z^2}, \tag{20}$$

together with the relation

$$\mathbf{j} = \pm \frac{H_0}{2\pi A_0}\, \omega = \pm \sqrt{\frac{\rho_0}{\pi \mu_e}}\, \omega. \tag{21}$$

13. Compressible Fluid—Vorticity and Current Density Propagation

In taking this view as a point of departure, we shall show that *in the case of a compressible medium, the components of* ω *and* $\mathbf{j}$, *in the direction of the field only, are propagated in Alfvén's manner.* Thus, surprisingly enough, the compressibility of a medium acts as a wave filter discriminating between components of vorticity—and current density—and passing only those directed along the (undisturbed) magnetic field. The proc is as follows. When compressibility is taken into account, the linearized system replacing (6) and (7) is

$$\rho_0 \frac{\partial \mathbf{v}}{\partial t} = -\operatorname{grad}\phi + \frac{\mu_e H_0}{4\pi} \frac{\partial \mathbf{h}}{\partial z} \tag{22}$$

$$\frac{\partial \mathbf{h}}{\partial t} = H_0 \frac{\partial \mathbf{v}}{\partial z} - \mathbf{H}_0 \operatorname{div}\mathbf{v} \tag{23}$$

where we set

$$\phi = p + \frac{\mu_e \mathbf{H}_0 \cdot \mathbf{h}}{4\pi} + \rho_0 \psi = p + \frac{\mu_e H_0 h_z}{4\pi} + \rho_0 \psi. \tag{24}$$

Equations (4) and (5) are replaced by

$$\frac{\partial \rho}{\partial t} + \rho_0 \operatorname{div} \mathbf{v} = 0 \tag{25}$$

$$\operatorname{div} \mathbf{v} = 0 \tag{26}$$

where ρ is the perturbation in density. We shall assume that

$$p = a_0^2 \rho \tag{27}$$

where a_0 is the ordinary sound speed in the absence of a magnetic field.

Taking the curl of terms of Eqs. (22) and (23) we obtain

$$2\rho_0 \frac{\partial \boldsymbol{\omega}}{\partial t} = \mu_e H_0 \frac{\partial \mathbf{j}}{\partial z} \tag{28}$$

$$4\pi \frac{\partial \mathbf{j}}{\partial t} = 2H_0 \frac{\partial \boldsymbol{\omega}}{\partial z} + \mathbf{H}_0 \times \operatorname{grad} \operatorname{div} \mathbf{v} \tag{29}$$

which imply important consequences, as will be shown.

(a) A Special Case of Propagation

In the case when ρ does not depend on x and y, then ω and $\mathbf{j}$ are propagated along the lines of force of the undisturbed magnetic field with velocities $\pm A_0$. In this case, by virtue of Eq. (25), Eq. (29) reduces to

$$2\pi \frac{\partial \mathbf{j}}{\partial t} = 2H_0 \frac{\partial \boldsymbol{\omega}}{\partial z}. \tag{30}$$

Combining Eqs. (28) and (30) we obtain

$$\frac{\partial^2 \boldsymbol{\omega}}{\partial t^2} = A_0^2 \frac{\partial^2 \boldsymbol{\omega}}{\partial z^2} \tag{31}$$

$$\frac{\partial^2 \mathbf{j}}{\partial t^2} = A_0^2 \frac{\partial^2 \mathbf{j}}{\partial z^2} \tag{32}$$

as in the case of an incompressible fluid.

(b) Propagation of z-Components of Vorticity and Current Density

Equations (28) and (29), when projected on the Oz axis, give

$$2\rho_0 \frac{\partial \omega_z}{\partial t} = \mu_e H_0 \frac{\partial j_z}{\partial z} \tag{33}$$

$$2\pi \frac{\partial j_z}{\partial t} = H_0 \frac{\partial \omega_z}{\partial z} . \tag{34}$$

Hence

$$\frac{\partial^2 \omega_z}{\partial t^2} = A_0^2 \frac{\partial^2 \omega_z}{\partial z^2} \tag{35}$$

$$\frac{\partial^2 j_z}{\partial t^2} = A_0^2 \frac{\partial^2 j_z}{\partial z^2} \tag{36}$$

and

$$j_z = \pm \frac{H_0}{2\pi A_0} \omega_z . \tag{37}$$

Thus the components of $\boldsymbol{\omega}$ and $\mathbf{j}$ along the lines of force (longitudinal components) are propagated in the opposite directions of the field with velocities $\pm A_0$.

The coupling relationship (37) between longitudinal components shows that

(i) it does not depend on the magnitude of the magnetic field present;

(ii) the vanishing of either component involves the vanishing of the other. This occurs when either quantity is zero initially.

Equation (36) is to be compared with Eq. (28) derived by Poeverlein [39] for the case of electromagnetic waves in a plasma with a strong magnetic field. In the latter equation the speed of light replaces Alfvén's velocity. This analogy deserves a closer investigation.

(c) Equations for the Transverse Components

We begin with the equations

$$\frac{\partial \omega_x}{\partial x} + \frac{\partial \omega_y}{\partial y} + \frac{\partial \omega_z}{\partial z} = 0 \tag{38}$$

$$\frac{\partial j_x}{\partial x} + \frac{\partial j_y}{\partial y} + \frac{\partial j_z}{\partial z} = 0. \tag{39}$$

The latter can be rewritten, by virtue of Eq. (37),

$$\frac{\partial j_x}{\partial x} + \frac{\partial j_y}{\partial y} \pm \frac{H_0}{2\pi A_0} \frac{\partial \omega_z}{\partial z} = 0. \tag{40}$$

Equations (38) and (40) indicate that

$$j_x = \pm \frac{H_0}{2\pi A_0} \omega_x + F(y, z, t); \tag{41}$$

$$F(y, z, 0) = 0, \tag{42}$$

$$j_y = \pm \frac{H_0}{2\pi A_0} \omega_y + G(x, z, t); \tag{43}$$

$$G(x, z, 0) = 0, \tag{44}$$

F and G being two arbitrary functions to be determined *a posteriori*. Substitution of the results (41) and (43) into Eqs. (28) and (29) yields

$$\frac{\partial \omega_x}{\partial t} \mp A_0 \frac{\partial \omega_x}{\partial z} = \frac{2\pi A_0^2}{H_0} \frac{\partial F}{\partial z} \tag{45}$$

$$\frac{\partial \omega_y}{\partial t} \mp A_0 \frac{\partial \omega_y}{\partial z} = \frac{2\pi A_0^2}{H_0} \frac{\partial G}{\partial z}, \tag{46}$$

and

$$\frac{\partial F}{\partial t} \pm A_0 \frac{\partial F}{\partial z} = \frac{H_0}{4\pi \rho_0} \frac{\partial^2 \rho}{\partial t\, \partial y} \tag{47}$$

$$\frac{\partial G}{\partial t} \pm A_0 \frac{\partial G}{\partial z} = - \frac{H_0}{4\pi\rho_0} \frac{\partial^2 \rho}{\partial t \, \partial x} . \tag{48}$$

Also, from Eqs. (41) and (43),

$$\frac{\partial j_x}{\partial t} \mp A_0 \frac{\partial j_x}{\partial z} = \frac{\partial F}{\partial t} \tag{49}$$

$$\frac{\partial j_y}{\partial t} \mp A_0 \frac{\partial j_y}{\partial z} = \frac{\partial G}{\partial t} . \tag{50}$$

Finally, combining Eqs. (45)-(49), we obtain

$$\frac{\partial^2 \omega_x}{\partial t^2} - A_0^2 \frac{\partial^2 \omega_x}{\partial z^2} = \frac{A_0^2}{2\rho_0} \frac{\partial^3 \rho}{\partial t \, \partial y \, \partial z} \tag{51}$$

$$\frac{\partial^2 \omega_y}{\partial t^2} - A_0^2 \frac{\partial^2 \omega_y}{\partial z^2} = - \frac{A_0^2}{2\rho_0} \frac{\partial^3 \rho}{\partial t \, \partial x \, \partial z} , \tag{52}$$

and

$$\frac{\partial^2 j_x}{\partial t^2} - A_0^2 \frac{\partial^2 j_x}{\partial z^2} = \frac{H_0}{4\pi\rho_0} \frac{\partial^3 \rho}{\partial t^2 \partial y} \tag{53}$$

$$\frac{\partial^2 j_y}{\partial t^2} - A_0^2 \frac{\partial^2 j_y}{\partial z^2} = - \frac{H_0}{4\pi\rho_0} \frac{\partial^3 \rho}{\partial t^2 \, \partial x} . \tag{54}$$

Equations (51)-(54) could, of course, be derived from Eqs. (28) and (29). The above analysis has, however, shown their fine structure. They yield, on elimination of ρ by cross-differentiation, Eqs. (35) and (36).

It is of interest to note that for an incompressible material, or in the case when ρ does not depend on x and y, the solutions of Eqs. (47) and (48) subject to initial conditions (42) and (44) are $F = G = 0$. This can easily be seen by taking their Laplace transforms with respect to time. In these cases only, the coupling relationship between $\mathbf{j}$ and $\boldsymbol{\omega}$ is identical to that resulting from Alfvén's theory [Eq. (21)]. In general, therefore, the vanishing of $\boldsymbol{\omega}$ will leave a current density in planes perpendicular to $\mathbf{H}_0$ and vice versa, as shown by formulas (41)-(44).

The combination of Eqs. (47) and (48) yields

$$\frac{\partial}{\partial y}\left(\frac{\partial F}{\partial t} \pm A_0 \frac{\partial F}{\partial z}\right) - \frac{\partial}{\partial x}\left(\frac{\partial G}{\partial t} \pm A_0 \frac{\partial G}{\partial z}\right) = \frac{H_0}{4\pi\rho_0} \frac{\partial}{\partial t}\left(\frac{\partial^2 \rho}{\partial x^2} + \frac{\partial^2 \rho}{\partial y^2}\right). \tag{55}$$

14. Wave-Motion Equation for the Hydrodynamic Density

Differentiation with respect to t of terms of Eq. (25) gives

$$\frac{\partial^2 \rho}{\partial t^2} = a_0^2 \nabla^2 \rho + \frac{A_0^2 \rho_0}{H_0} \nabla^2 h_z \tag{56}$$

where Eqs. (22) and (27) have been used, and the gravitational potential has been omitted.

Now

$$\nabla^2 h_z = 4\pi \left(\frac{\partial j_x}{\partial y} - \frac{\partial j_y}{\partial x}\right). \tag{57}$$

Using Eqs. (41) and (43) this can be rewritten

$$\nabla^2 h_z = \pm \frac{2H_0}{A_0}\left(\frac{\partial \omega_x}{\partial y} - \frac{\partial \omega_y}{\partial x}\right) + 4\pi\left(\frac{\partial F}{\partial y} - \frac{\partial G}{\partial x}\right). \tag{58}$$

Therefore

$$\frac{\partial^2 \rho}{\partial t^2} = a_0^2 \nabla^2 \rho \pm 2A_0 \rho_0 \left(\frac{\partial \omega_x}{\partial y} - \frac{\partial \omega_y}{\partial x}\right) + \frac{4\pi A_0^2 \rho_0}{H_0}\left(\frac{\partial F}{\partial y} - \frac{\partial G}{\partial x}\right). \tag{59}$$

The functions F and G can now be eliminated to obtain an equation for ρ alone. It is more convenient, however, to obtain this equation by starting with Eq. (56) and eliminating j_x, j_y between Eqs. (57) and (53), (54). The result is

$$\frac{\partial^2}{\partial t^2}\left(\frac{\partial^2\rho}{\partial t^2} - a_0^2\nabla^2\rho\right) - A_0^2\frac{\partial^2}{\partial z^2}\left(\frac{\partial^2\rho}{\partial t^2} - a_0^2\nabla^2\rho\right)$$

$$= A_0^2\frac{\partial^2}{\partial t^2}\left(\frac{\partial^2\rho}{\partial x^2} + \frac{\partial^2\rho}{\partial y^2}\right). \tag{60}$$

We may now ask if it be possible to satisfy both Eqs. (60) and

$$\frac{\partial^2\rho}{\partial t^2} - a_0^2\nabla^2\rho = 0. \tag{61}$$

Then, Eq. (60) requires *(under special initial conditions)*

$$\frac{\partial^2\rho}{\partial x^2} + \frac{\partial^2\rho}{\partial y^2} = 0, \tag{62}$$

and hence Eq. (61) reduces to

$$\frac{\partial^2\rho}{\partial t^2} - a_0^2\frac{\partial^2\rho}{\partial z^2} = 0 \tag{63}$$

which admits a solution of the form

$$\rho \cong \exp[i(\omega t - \gamma z)], \tag{64}$$

provided that

$$\gamma^2 = \frac{\omega^2}{a_0^2}. \tag{65}$$

We can easily verify that (64) is a particular solution of Eq. (60) under condition (65). Thus, *sound waves appear possible in a conducting fluid penetrated by a uniform magnetic field, with this great difference: that they do not spread out three-dimensionally as in ordinary acoustics. Instead, they propagate (without attenuation) one-dimensionally, along the magnetic lines of force.* It is also interesting to note that in contrast to mh waves this propagation does not depend on the magnitude of the magnetic field present. We shall give a more detailed analysis of Eq. (60) later. We notice now that this equation can be rewritten

$$\frac{\partial^2}{\partial t^2}\left(\frac{\partial^2\rho}{\partial t^2} - (a_0^2 + A_0^2)\nabla^2\rho\right) + a_0^2A_0^2\,\frac{\partial^2}{\partial z^2}\,\nabla^2\rho = 0. \tag{66}$$

Differentiation with respect to t of terms of Eq. (66) yields an equation given by Lighthill [32] for div $\mathbf{v}$. For time harmonic dependence, one finds a result given earlier by Baños [5, p. 352].

Now, if a_0 and A_0 are small enough to neglect their product, Eq. (66) simplifies, under special initial conditions, to

$$\frac{\partial^2\rho}{\partial t^2} - (a_0^2 + A_0^2)\nabla^2\rho = 0, \tag{67}$$

which is the wave equation for ρ encountered in acoustics, but when the square of the speed of sound a_0^2 is replaced by $a_0^2 + A_0^2$.

15. Wave-Motion Equations for the Transverse Components

Let us return to Eqs. (51)-(54) and improve our results. Elimination of ρ between Eqs. (51)-(52) and (66) gives

$$\begin{aligned}\Bigg\{&\frac{\partial^2}{\partial t^2}\left[\frac{\partial^2}{\partial t^2}\left(\frac{\partial^2}{\partial t^2} - (a_0^2 + A_0^2)\nabla^2\right) + a_0^2A_0^2\,\frac{\partial^2}{\partial z^2}\,\nabla^2\right]\\ &- A_0^2\,\frac{\partial^2}{\partial z^2}\left[\frac{\partial^2}{\partial t^2}\left(\frac{\partial^2}{\partial t^2} - (a_0^2 + A_0^2)\nabla^2\right) + a_0^2A_0^2\,\frac{\partial^2}{\partial z^2}\,\nabla^2\right]\Bigg\}\omega_x = 0,\end{aligned} \tag{68}$$

$$\begin{aligned}\Bigg\{&\frac{\partial^2}{\partial t^2}\left[\frac{\partial^2}{\partial t^2}\left(\frac{\partial^2}{\partial t^2} - (a_0^2 + A_0^2)\nabla^2\right) + a_0^2A_0^2\,\frac{\partial^2}{\partial z^2}\,\nabla^2\right]\\ &- A_0^2\,\frac{\partial^2}{\partial z^2}\left[\frac{\partial^2}{\partial t^2}\left(\frac{\partial^2}{\partial t^2} - (a_0^2 + A_0^2)\nabla^2\right) + a_0^2A_0^2\,\frac{\partial^2}{\partial z^2}\,\nabla^2\right]\Bigg\}\omega_y = 0.\end{aligned} \tag{69}$$

Similarly, elimination of ρ between Eqs. (53)-(54) and (66) gives

$$\left\{ \frac{\partial^2}{\partial t^2} \left[\frac{\partial^2}{\partial t^2} \left(\frac{\partial^2}{\partial t^2} - (a_0^2 + A_0^2) \nabla^2 \right) + a_0^2 A_0^2 \frac{\partial^2}{\partial z^2} \nabla^2 \right] \right.$$

$$\left. - A_0^2 \frac{\partial^2}{\partial z^2} \left[\frac{\partial^2}{\partial t^2} \left(\frac{\partial^2}{\partial t^2} - (a_0^2 + A_0^2) \nabla^2 \right) + a_0^2 A_0^2 \frac{\partial^2}{\partial z^2} \nabla^2 \right] \right\} j_x = 0 , \tag{70}$$

$$\left\{ \frac{\partial^2}{\partial t^2} \left[\frac{\partial^2}{\partial t^2} \left(\frac{\partial^2}{\partial t^2} - (a_0^2 + A_0^2) \nabla^2 \right) + a_0^2 A_0^2 \frac{\partial^2}{\partial z^2} \nabla^2 \right] \right.$$

$$\left. - A_0^2 \frac{\partial^2}{\partial z^2} \left[\frac{\partial^2}{\partial t^2} \left(\frac{\partial^2}{\partial t^2} - (a_0^2 + A_0^2) \nabla^2 \right) + a_0^2 A_0^2 \frac{\partial^2}{\partial z^2} \nabla^2 \right] \right\} j_y = 0 . \tag{71}$$

Equations (68)-(71) show that the quantities

$$\left[\frac{\partial^2}{\partial t^2} \left(\frac{\partial^2}{\partial t^2} - (a_0^2 + A_0^2) \nabla^2 \right) + a_0^2 A_0^2 \frac{\partial^2}{\partial z^2} \nabla^2 \right]^{\omega_x , \omega_y}_{j_x , j_y} \tag{72}$$

are propagated along magnetic lines of force at Alfvén velocity A_0. These quantities are identically zero if they are zero initially. Under this condition, Eqs. (68)-(71) reduce to fourth-order equations of the same type as Eq. (66). As it has already been observed by Lighthill [32] three-dimensional equations of type (66) lead to spherical attenuation. Any local disturbance, therefore, does in part get propagated one-dimensionally, the part in question being the component of vorticity along the magnetic lines of force, and current density, as coupled with vorticity by Eq. (37). Far away, where $\operatorname{div} \mathbf{v}$, ω_x, ω_y, j_x, j_y have been attenuated to values small compared with ω_z and j_z, the disturbance becomes a purely two-dimensional, solenoidal, vortex motion. Instead of being convected with the fluid in the way that such a disturbance would be in an ordinary conservative field, it propagates (in both directions) along the magnetic lines of force. Hence, only far away can one properly speak in this case of *transverse waves*, since then both $\mathbf{v}$ and $\mathbf{h}$ become perpendicular to the direction of propagation.

16. General Coupling Relationship between Vorticity and Current Density

The coupling relationship between vorticity and current density given by formulas (37) and (41)-(44) can be generalized, as suggested to us by Walén (personal communication), in the following way. One can write, generally,

$$\mathbf{j} = \pm \frac{H_0}{2\pi A_0} \omega + \mathbf{H}_0 \times \operatorname{grad} \phi , \tag{73}$$

for relation (37) is satisfied and

$$\operatorname{div} \mathbf{j} = \operatorname{div} (\mathbf{H}_0 \times \operatorname{grad} \phi) = - \mathbf{H}_0 \operatorname{curl} \operatorname{grad} \phi = 0 . \tag{74}$$

To determine ρ, we substitute (73) into Eqs. (28) and (29). This gives

$$\mathbf{H}_0 \times \operatorname{grad} \left(\frac{\partial \phi}{\partial t} \pm A_0 \frac{\partial \phi}{\partial z} + \frac{1}{4\pi\rho_0} \frac{\partial \rho}{\partial t} \right) = 0. \tag{75}$$

Hence, ρ satisfies

$$\frac{\partial \phi}{\partial t} \pm A_0 \frac{\partial \phi}{\partial z} + \frac{1}{4\pi\rho_0} \frac{\partial \rho}{\partial t} = 0. \tag{76}$$

This is Walén's result. Now, we can go further and observe that elimination of ρ between Eqs. (66) and (76) gives

$$\frac{\partial}{\partial t} \left[\frac{\partial^2}{\partial t^2} \left(\frac{\partial^2 \phi}{\partial t^2} - (a_0^2 + A_0^2) \nabla^2 \phi \right) + a_0^2 A_0^2 \frac{\partial^2}{\partial z^2} \nabla^2 \phi \right]$$

$$\pm A_0 \frac{\partial}{\partial z} \left[\frac{\partial^2}{\partial t^2} \left(\frac{\partial^2 \phi}{\partial t^2} - (a_0^2 + A_0^2) \nabla^2 \phi \right) + a_0^2 A_0^2 \frac{\partial^2}{\partial z^2} \nabla^2 \phi \right] = 0. \tag{77}$$

Hence, ϕ verifies

$$\left\{ \frac{\partial^2}{\partial t^2} \left[\frac{\partial^2}{\partial t^2} \left(\frac{\partial^2}{\partial t^2} - (a_0^2 + A_0^2) \nabla^2 \right) + a_0^2 A_0^2 \frac{\partial^2}{\partial z^2} \nabla^2 \right] \right.$$

$$\left. - A_0^2 \frac{\partial^2}{\partial z^2} \left[\frac{\partial^2}{\partial t^2} \left(\frac{\partial^2}{\partial t^2} - (a_0^2 + A_0^2) \nabla^2 \right) + a_0^2 A_0^2 \frac{\partial^2}{\partial z^2} \right] \right\} \phi = 0 , \tag{78}$$

as one would have expected by virtue of Eqs. (68)-(71). Thus, far away from a source of disturbances the function ϕ will get attenuated and the coupling (37) will then replace that given by Eq. (73).

17. Wave-Motion Equations for e_{ij} and γ_{ij}

We shall now study how the variations produced in the rate of deformation e_{ij} and γ_{ij} [see formula (102), Part I]

$$\gamma_{ij} = \frac{1}{2}\left(\frac{\partial h_i}{\partial x_j} + \frac{\partial h_j}{\partial x_i}\right) \tag{79}$$

are propagated.

Lighthill [32] was the first to give a wave equation for the e_{33} component of e_{ij} which, as observed by him, was identical with that satisfied by $\operatorname{div} \mathbf{v}$. Earlier, Grad [25] gave similar equations (coupled) for the longitudinal components of material velocity and magnetic induction. But until this writer's paper [11], there had been no systematic attempt to formulate wave equations for e_{ij} and γ_{ij}. These are precisely the equations which will be derived in this section.

When e_{ij} and γ_{ij} are introduced in Eqs. (22) and (23) (we omit here the gravitational potential), these equations become

$$\rho_0 \frac{\partial e_{11}}{\partial t} = - a_0^2 \frac{\partial^2 \rho}{\partial x^2} + \mu_e H_0 \frac{\partial j_y}{\partial x},$$

$$\rho_0 \frac{\partial e_{22}}{\partial t} = - a_0^2 \frac{\partial^2 \rho}{\partial t^2} - \mu_e H_0 \frac{\partial j_x}{\partial y},$$

$$\rho_0 \frac{\partial e_{33}}{\partial t} = - a_0^2 \frac{\partial^2 \rho}{\partial z^2},$$

$$\rho_0 \frac{\partial e_{23}}{\partial t} = - a_0^2 \frac{\partial^2 \rho}{\partial y \partial z} - \frac{\mu_e H_0}{2} \frac{\partial j_x}{\partial z},$$

$$\rho_0 \frac{\partial e_{31}}{\partial t} = - a_0^2 \frac{\partial^2 \rho}{\partial z \partial x} + \frac{\mu_e H_0}{2} \frac{\partial j_y}{\partial z}, \tag{80}$$

(Cont.)

$$\rho_0 \frac{\partial e_{12}}{\partial t} = - a_0^2 \frac{\partial^2 \rho}{\partial x \partial y} + \frac{\mu_e H_0}{2} \left(\frac{\partial j_y}{\partial y} - \frac{\partial j_x}{\partial x} \right); \tag{80}$$

and

$$\begin{aligned}
\frac{\partial \gamma_{11}}{\partial t} &= H_0 \frac{\partial e_{11}}{\partial z}, \\
\frac{\partial \gamma_{22}}{\partial t} &= H_0 \frac{\partial e_{22}}{\partial z}, \\
\frac{\partial \gamma_{33}}{\partial t} &= \frac{H_0}{\rho_0} \frac{\partial^2 \rho}{\partial t \, \partial z} + H_0 \frac{\partial e_{33}}{\partial z}, \\
\frac{\partial \gamma_{23}}{\partial t} &= \frac{H_0}{2\rho_0} \frac{\partial^2 \rho}{\partial t \, \partial y} + H_0 \frac{\partial e_{23}}{\partial z}, \\
\frac{\partial \gamma_{31}}{\partial t} &= \frac{H_0}{2\rho_0} \frac{\partial^2 \rho}{\partial t \, \partial x} + H_0 \frac{\partial e_{31}}{\partial z}, \\
\frac{\partial \gamma_{12}}{\partial t} &= H_0 \frac{\partial e_{12}}{\partial z}.
\end{aligned} \tag{81}$$

These equations are highly asymmetrical and show perhaps more plainly than Eqs. (22) and (23) the strongly anisotropic character of wave motion of the electrically conducting fluid. Using the results obtained in Sections 14 and 15, however, the conclusions to be drawn are surprisingly simple. First, it follows at once by virtue of Eq. (66) that e_{33} satisfies

$$\frac{\partial^2}{\partial t^2} \left(\frac{\partial^2 e_{33}}{\partial t^2} - (a_0^2 + A_0^2) \nabla^2 e_{33} \right) + a_0^2 A_0^2 \frac{\partial^2}{\partial z^2} \nabla^2 e_{33} = 0, \tag{82}$$

a result obtained earlier by Lighthill ([32], p. 401, Eq. 17). We may add that the same equation is also satisfied by γ_{33} as seen by inspection of the third equation in (81).

By successive eliminations, we then find that all other components satisfy a sixth-order differential equation of the type as that encountered in Section 15. We have, therefore,

$$\left\{\frac{\partial^2}{\partial t^2}\left[\frac{\partial^2}{\partial t^2}\left(\frac{\partial^2}{\partial t^2} - (a_0^2 + A_0^2)\nabla^2\right) + a_0^2 A_0^2 \frac{\partial^2}{\partial z^2}\nabla^2\right]\right.$$

$$\left. - A_0^2 \frac{\partial^2}{\partial z^2}\left[\frac{\partial^2}{\partial t^2}\left(\frac{\partial^2}{\partial t^2} - (a_0^2 + A_0^2)\nabla^2\right) + a_0^2 A_0^2 \frac{\partial^2}{\partial z^2}\nabla^2\right]\right\} \begin{matrix} e'_{ij} \\ \gamma'_{ij} \end{matrix} = 0 \tag{83}$$

where we use the prime in order to exclude the values $i = j = 3$ of suffixes; for $i = j = 3$, we have Eq. (82).

18. Plane-Wave Solutions and Wave-Number Surface

Consider plane-wave solutions

$$\begin{matrix} e'_{ij} \\ \gamma'_{ij} \end{matrix} = \begin{matrix} (e_{ij})'_0 \\ (\gamma_{ij})'_0 \end{matrix} \exp\left[i(\omega t - \alpha x - \beta y - \gamma z)\right] \tag{84}$$

of Eq. (83). Substitution of (84) into (83) yields the following wave-number surface (locus of points α, β, γ for given ω):

$$(\omega^2 - A_0^2\gamma^2)\left[(\omega^2 - a_0^2\gamma^2)(\omega^2 - A_0^2\gamma^2) - (\alpha^2 + \beta^2)\left((a_0^2 + A_0^2)\omega^2 - a_0^2 A_0^2\gamma^2\right)\right] = 0. \tag{85}$$

Hence:

(a) either $\gamma = \pm\omega/A_0$ and then e'_{ij}, γ'_{ij} are propagated solely along magnetic lines of force without attenuation at the Alfvén velocity A_0;

(b) or $\gamma \neq \pm\omega/A_0$, and then

$$\alpha^2 + \beta^2 = \frac{(\omega^2 - a_0^2\gamma^2)(\omega^2 - A_0^2\gamma^2)}{(a_0^2 + A_0^2)\omega^2 - a_0^2 A_0^2\gamma^2}, \tag{86}$$

which represents the wave-number surface given by Lighthill [32]. This is a surface of revolution, with its axis the γ-axis (which represents the undisturbed magnetic field) and a plane of symmetry $\gamma = 0$. It is in three sheets, one an ovoid of major axis ω/a_0 and minor axis $\omega/\sqrt{a_0^2 + A_0^2}$, and the other two consisting of the near planes $\gamma = \pm\omega\sqrt{a_0^{-2} + A_0^{-2}}$ disturbed by bumps near the γ-axis, which reduce the distance between the disturbed surfaces to twice ω/A_0 (see Lighthill [32], p. 416). Under (b) we distinguish two subcases:

(b.1) $\gamma = \pm\omega/a_0$ and then e'_{ij}, γ'_{ij} are propagated one-dimensionally along magnetic lines of force, at the sound speed a_0; in this case $\alpha = \beta = 0$, that is, the ovoid sheet of (86) collapses into the γ-axis.

(b.2) $\gamma \neq \pm\omega/a_0$; by rewriting Eq. (86) as

$$\alpha^2 + \beta^2 = \frac{1}{A_0^2}\,\frac{(\omega^2 - a_0^2\gamma^2)(\omega^2 - A_0^2\gamma^2)}{(1 + a_0^2/A_0^2)\,\omega^2 - a_0^2\gamma^2}, \tag{87}$$

or

$$\alpha^2 + \beta^2 = \frac{1}{a_0^2}\,\frac{(\omega^2 - a_0^2\gamma^2)(\omega^2 - A_0^2\gamma^2)}{(1 + A_0^2/a_0^2)\,\omega^2 - A_0^2\gamma^2}, \tag{88}$$

two limiting cases, considered in slightly different form by Baños [5], Grad [25] and Lighthill [32], are readily revealed:

(b.2.1) $A_0 \gg a_0$, when Eq. (87) reduces approximately to

$$\alpha^2 + \beta^2 + \gamma^2 = \frac{\omega^2}{A_0^2}, \tag{89}$$

and we have propagation in *all* directions with nearly equal velocity A_0;

(b.2.2) $A_0 \ll a_0$ when Eq. (88) reduces approximately to

$$\alpha^2 + \beta^2 + \gamma^2 = \frac{\omega^2}{a_0^2}, \tag{90}$$

and we have isotropic tri-dimensional propagation with velocity nearly a_0.

The singular case $\gamma = \pm\omega\sqrt{a_0^{-2} + A_0^{-2}}$ when $\alpha^2 + \beta^2 = \infty$ requires more careful analysis and is omitted here.

19. Compressibility Effect on Velocity and Magnetic Field

The wave equations for e_{ij} and γ_{ij} may also be derived by the following brief analysis, which re-emphasizes the compressibility effect of discrimination between longitudinal and transverse components.

The z-component of material velocity satisfies

$$\rho_0 \frac{\partial v_z}{\partial t} = -a_0^2 \frac{\partial \rho}{\partial z} . \tag{91}$$

Hence, by virtue of Eq. (66), we have at once

$$\left[\frac{\partial^2}{\partial t^2}\left(\frac{\partial^2}{\partial t^2} - (a_0^2 + A_0^2)\nabla^2\right) + a_0^2 A_0^2 \frac{\partial^2}{\partial z^2}\right] v_z = 0. \tag{92}$$

On the other hand,

$$\frac{\partial h_z}{\partial t} = H_0 \frac{\partial v_z}{\partial z} - H_0 \operatorname{div} \mathbf{v}. \tag{93}$$

Since both v_z and $\operatorname{div} \mathbf{v}$ satisfy (92), it follows that

$$\left[\frac{\partial^2}{\partial t^2}\left(\frac{\partial^2}{\partial t^2} - (a_0^2 + A_0^2)\nabla^2\right) + a_0^2 A_0^2 \frac{\partial^2}{\partial z^2}\nabla^2\right] h_z = 0. \tag{94}$$

Combining equations for the transverse components in Eqs. (22) (where the gravitational potential is omitted) and (23) with Eqs. (66), (92), and (94), we obtain

$$\left\{\frac{\partial^2}{\partial t^2}\left[\frac{\partial^2}{\partial t^2}\left(\frac{\partial^2}{\partial t^2} - (a_0^2 + A_0^2)\nabla^2\right) + a_0^2 A_0^2 \frac{\partial^2}{\partial z^2}\nabla^2\right]\right.$$

$$\left. - A_0^2 \frac{\partial^2}{\partial z^2}\left[\frac{\partial^2}{\partial t^2}\left(\frac{\partial^2}{\partial t^2} - (a_0^2 + A_0^2)\nabla^2\right) + a_0^2 A_0^2 \frac{\partial^2}{\partial z^2}\nabla^2\right]\right\} \begin{matrix} v_x, v_y \\ h_x, h_y \end{matrix} = 0. \tag{95}$$

The wave-motion equations for e_{ij} and γ_{ij} become obvious.

III. Dissipative Effects in Magnetohydrodynamic Waves. The Effect of Finite Electrical Conductivity

20. Preliminaries

The theory presented thus far ignored dissipative effects. We shall now briefly investigate the effects of electrical resistance and viscosity, assuming that these quantities are constant. Also, we shall take into account here the effect of the compressibility of the material. For the case of an incompressible fluid, but when ν and ν_e are variable with respect to position, we refer to the excellent analysis of Chandrasekhar ([15], see pp. 531-576). We shall see that the phenomena involved become highly complex as the equations governing them are partial differential equations of order higher than second and of a mixed type: *hyperbolic-parabolic,* which are very little written about.

The dissipative effects have been studied by several authors. Besides Chandrasekhar's work already mentioned (where more references are given) we especially note here the papers of Baños [4, 5], Carstoiu [12, 14], Lehnert [31], Ludford [34], Resler and McCune [40], and Resler [41].

It is known that one can neglect the effect of either frictional term introduced by ν and ν_e depending on how small or large the ratio ν/ν_e may be. Nonetheless, the analysis presented in the pertinent literature (see for instance, Cowling [16], p. 38) is rather intuitive and little supported. In Part IV, where the combined effects of electrical conductivity and viscosity are presented, the ratio of these quantities and how they are interconnected will be outlined. In Part III we shall especially study how a finite electrical conductivity *alone* effects magnetohydrodynamic wave propagation. This is usually the case in cosmic electrodynamics, except in the extremely tenuous matter in interstellar space.

21. Fundamental Equations

The relevant perturbation equations are (Carstoiu [12])

$$\rho_0 \frac{\partial \mathbf{v}}{\partial t} = -\operatorname{grad} \phi + \frac{\mu_e H_0}{4\pi} \frac{\partial \mathbf{h}}{\partial z} \tag{1}$$

$$\frac{\partial \mathbf{h}}{\partial t} = H_0 \frac{\partial \mathbf{v}}{\partial z} - \mathbf{H}_0 \operatorname{div} \mathbf{v} + \nu_e \nabla^2 \mathbf{h} \tag{2}$$

$$\frac{\partial \rho}{\partial t} + \rho_0 \operatorname{div} \mathbf{v} = 0 \tag{3}$$

$$\operatorname{div} \mathbf{h} = 0 \tag{4}$$

where $\phi = a_0^2 \rho + (\mu_e H_0 h_z)/4\pi$.

Taking the curl of terms of Eqs. (1) and (2), one obtains

$$\frac{\partial \omega}{\partial t} = \frac{\mu_e H_0}{2\rho_0} \frac{\partial \mathbf{j}}{\partial z} \tag{5}$$

$$\frac{\partial \mathbf{j}}{\partial t} = \frac{H_0}{2\pi} \frac{\partial \omega}{\partial z} - \frac{\mathbf{H}_0}{4\pi\rho_0} \times \operatorname{grad} \frac{\partial \rho}{\partial t} + \nu_e \nabla^2 \mathbf{j} \tag{6}$$

where Eq. (3) has been used.

It follows that the longitudinal components ω_z and j_z of vorticity and current density verify the following equations:

$$\frac{\partial \omega_z}{\partial t} = \frac{\mu_e H_0}{2\rho_0} \frac{\partial j_z}{\partial z} \tag{7}$$

$$\frac{\partial j_z}{\partial t} = \frac{H_0}{2\pi} \frac{\partial \omega_z}{\partial z} + \nu_e \nabla^2 j_z \,. \tag{8}$$

Cross-differentiation of these equations yields

$$\frac{\partial^2 \omega_z}{\partial t^2} - A_0^2 \frac{\partial^2 \omega_z}{\partial z^2} - \nu_e \frac{\partial}{\partial t} \nabla^2 \omega_z = 0 \tag{9}$$

$$\frac{\partial^2 j_z}{\partial t^2} - A_0^2 \frac{\partial^2 j_z}{\partial z^2} - \nu_e \frac{\partial}{\partial t} \nabla^2 j_z = 0 \,. \tag{10}$$

We see that, as for the case of an infinite conductivity, the longitudinal components of $\boldsymbol{\omega}$ and $\mathbf{j}$ ignore the oscillations of the density ρ. Now, however, the diffusion effects become apparent. We can roughly describe the motion as the sum of progressive waves traveling with velocities $\pm A_0$ in the z-direction, but with the disturbances diffusing outward as they travel. Now, while the diffusion effects may not be important in most cosmic problems, they are very important in laboratory experiments.

It may also be noted that the oscillations of the transverse components of $\boldsymbol{\omega}$ and $\mathbf{j}$ are coupled to those of density ρ. For the oscillations of these components satisfy

$$\frac{\partial^2\omega_x}{\partial t^2} - A_0^2 \frac{\partial^2\omega_x}{\partial z^2} - \nu_e \frac{\partial}{\partial t} \nabla^2\omega_x = \frac{A_0^2}{2\rho_0} \frac{\partial^3\rho}{\partial t\, \partial y\, \partial z} \tag{11}$$

$$\frac{\partial^2\omega_y}{\partial t^2} - A_0^2 \frac{\partial^2\omega_y}{\partial z^2} - \nu_e \frac{\partial}{\partial t} \nabla^2\omega_y = -\frac{A_0^2}{2\rho_0} \frac{\partial^3\rho}{\partial t\, \partial x\, \partial z}, \tag{12}$$

and

$$\frac{\partial^2 j_x}{\partial t^2} - A_0^2 \frac{\partial^2 j_x}{\partial z^2} - \nu_e \frac{\partial}{\partial t} \nabla^2 j_x = \frac{A_0^2}{\mu_e H_0} \frac{\partial^3\rho}{\partial t^2\, \partial y} \tag{13}$$

$$\frac{\partial^2 j_y}{\partial t^2} - A_0^2 \frac{\partial^2 j_y}{\partial z^2} - \nu_e \frac{\partial}{\partial t} \nabla^2 j_y = -\frac{A_0^2}{\mu_e H_0} \frac{\partial^3\rho}{\partial t^2\, \partial x}. \tag{14}$$

These equations show plainly that only for an incompressible material do the transverse components satisfy the same equations (9) and (10).

22. Wave-Motion Equation for the Hydrodynamic Density

The density ρ continues to satisfy the equation [see Eqs. (56) and (57) in Part II]

$$\frac{\partial^2\rho}{\partial t^2} - a_0^2 \nabla^2\rho = \mu_e H_0 \left(\frac{\partial j_x}{\partial y} - \frac{\partial j_y}{\partial x}\right), \tag{15}$$

but now the elimination of j_x and j_y between Eqs. (13)-(15) will produce a far more complicated equation for ρ than Eq. (66) of Part II. The result of this elimination is (Carstoiu [12])

$$\left\{\frac{\partial^2}{\partial t^2}\left[\frac{\partial^2}{\partial t^2} - \left(a_0^2 + A_0^2 + \nu_e \frac{\partial}{\partial t}\right)\nabla^2\right]\right.$$

$$\left. + a_0^2 \nabla^2 \left(A_0^2 \frac{\partial^2}{\partial z^2} + \nu_e \frac{\partial}{\partial t} \nabla^2\right)\right\}\rho = 0. \tag{16}$$

It is remarkable, however, that this equation admits a particular solution of the same type as that studied in Section 14. For, let us try a plane-wave solution of the form

$$\rho \cong \exp\left[i(\omega t - \alpha x - \beta y - \gamma z)\right] , \tag{17}$$

when all quantities ω, α, s, γ are assumed to be real.

Substitution of (17) into (16) gives

$$\omega^4 - (a_0^2 + A_0^2)(\alpha^2 + \beta^2 + \gamma^2)\,\omega^2 + a_0^2 A_0^2(\alpha^2 + \beta^2 + \gamma^2)\,\gamma^2 = 0 \tag{18}$$

together with

$$\alpha^2 + \beta^2 + \gamma^2 = \frac{\omega^2}{a_0^2} . \tag{19}$$

Equation (18) represents the Lighthill's wave number surface [see Eq. (86), Part II] while (19) gives a sphere of radius ω/a_0. By virtue of Eq. (19), Eq. (18) simplifies to

$$\gamma^2 = \frac{\omega^2}{a_0^2} , \tag{20}$$

and hence, by (19),

$$\alpha = \beta = 0 \tag{21}$$

which gives the same solution previously obtained when $\sigma \to \infty$.

This result is extremely intriguing and indicates the possibility of one-dimensional sound wave propagation, along the magnetic lines of force, in a fluid with finite electrical conductivity. It is also observed that the value of this conductivity does not appear in our equations. This problem requires more investigation and is left open here for further study.

23. Wave-Motion Equations for the Transverse Components of ω and $\mathbf{j}$, for the Velocity, and for the Magnetic Field

We can proceed, following the method used in Part II, to eliminate the density ρ between Eqs. (11)-(14) and (16) in order to obtain wave-motion equations for the transverse components of ω and $\mathbf{j}$. In view of the length of these equations, it now becomes necessary to use some appropriated operators in order to abbreviate our equations formally. We put

$$\mathbf{P} \equiv \frac{\partial^2}{\partial t^2} - A_0^2 \frac{\partial^2}{\partial z^2} - \nu_e \frac{\partial}{\partial t} \nabla^2 \tag{22}$$

and

$$Q \equiv \frac{\partial^2}{\partial t^2}\left[\frac{\partial^2}{\partial t^2} - \left(a_0^2 + A_0^2 + \nu_e \frac{\partial}{\partial t}\right)\nabla^2\right] + a_0^2 \nabla^2 \left(A_0^2 \frac{\partial^2}{\partial z^2} + \nu_e \frac{\partial}{\partial t} \nabla^2\right) . \tag{23}$$

Using these operators we then immediately obtain

$$\mathbf{P}Q\omega_x \equiv \mathbf{P}Q\omega_y = \mathbf{P}Qj_x = \mathbf{P}Qj_y = 0 \tag{24}$$

where obviously we can permute the order of operators $\mathbf{P}$ and Q.

We now return to Eqs. (1) and (2) and transform them using the results already obtained. We shall also use the following operator:

$$D \equiv \frac{\partial}{\partial t} - \nu_e \nabla^2 . \tag{25}$$

The following equations for the velocity $\mathbf{v}$ and the magnetic field $\mathbf{h}$ are then obtained:

$$\mathbf{P}Qv_x = \mathbf{P}Qv_y = 0 \tag{26}$$

$$Qv_z = 0 , \tag{27}$$

and

$$DPQh_x = DPQh_y = 0 \tag{28}$$

$$DQh_z = 0. \tag{29}$$

Equations (26) and (27) imply similar equations for the rate of deformation e_{ij}. Equations (28) and (29) yield like equations for γ_{ij}. It is also observed that $\operatorname{div} \mathbf{v}$ satisfies an equation of type (16).

IV. Dissipative Effects in Magnetohydrodynamic Waves. The Effect of Viscosity. The Combined Effects of Finite Electrical Conductivity and Viscosity

24. The Effect of Viscosity

We can now alternatively study how the viscosity alone will affect magnetohydrodynamic wave propagation. This will not, however, be necessary, except perhaps as a mathematical exercise. Indeed, when one specifies the disturbance by means of vorticity and current density, then, curiously enough, the effect of the viscosity is identical to that of electrical conductivity; this means that governing equations for vorticity and current density are identical to those given in Part III [Eqs. (9)-(14)] except that the *coefficient of magnetic viscosity* $\nu_e = (4\pi\mu_e\sigma)^{-1}$ is to be replaced by the *coefficient of kinematic viscosity* $\nu = \mu/\rho_0$. The radical departure between the two effects—electrical conductivity and viscosity—appears in the equation governing the density ρ. This will be analyzed in the following sections.

25. Combined Effects of Electrical Conductivity and Viscosity

If, as in previous analyses, the z-axis is taken in the direction of the field $\mathbf{H}_0$, the equations giving the variations in $\mathbf{v}$, $\mathbf{h}$, and ρ are

$$\rho_0 \frac{\partial \mathbf{v}}{\partial t} = -\operatorname{grad}\left[\phi - (\lambda + \mu)\operatorname{div}\mathbf{v}\right] + \frac{\mu_e H_0}{4\pi}\frac{\partial \mathbf{h}}{\partial z} + \mu\nabla^2\mathbf{v} \tag{1}$$

$$\frac{\partial \mathbf{h}}{\partial t} = H_0 \frac{\partial \mathbf{v}}{\partial z} - \mathbf{H}_0 \operatorname{div} \mathbf{v} + \nu_e \nabla^2 \mathbf{h} \tag{2}$$

$$\frac{\partial \rho}{\partial t} + \rho_0 \operatorname{div} \mathbf{v} = 0 \tag{3}$$

$$\operatorname{div} \mathbf{h} = 0 \tag{4}$$

where besides notations already used, λ and μ are the two coefficients of viscosity, namely, *bulk* and *shear* viscosity, respectively, which we here consider as constants depending on the nature of the fluid and on its physical state.

It is generally admitted (see for instance Chandrasekhar [15]) that

$$3\lambda + 2\mu = 0. \tag{5}$$

This relation has been used in Eq. (8) of Part I. We refer for further discussion about condition (5) to Villat's book [50, pp. 66-77] and use here, for more generality, Eq. (1). Again, the gravitational potential will be omitted in ϕ.

Taking the curl of terms of Eqs. (1) and (2) we now obtain

$$\frac{\partial \boldsymbol{\omega}}{\partial t} = \frac{\mu_e H_0}{2\rho_0} \frac{\partial \mathbf{j}}{\partial z} + \nu \nabla^2 \boldsymbol{\omega} \tag{6}$$

$$\frac{\partial \mathbf{j}}{\partial t} = \frac{H_0}{2\pi} \frac{\partial \boldsymbol{\omega}}{\partial z} - \frac{\mathbf{H}_0}{4\pi\rho_0} \times \operatorname{grad} \frac{\partial \rho}{\partial t} + \nu_e \nabla^2 \mathbf{j}. \tag{7}$$

It follows that the longitudinal components ω_z and j_z of vorticity and current density satisfy the following equations:

$$\frac{\partial^2 \omega_z}{\partial t^2} - A_0^2 \frac{\partial^2 \omega_z}{\partial z^2} - \nabla^2 \left[(\nu + \nu_e) \frac{\partial \omega_z}{\partial t} - \nu \nu_e \nabla^2 \omega_z \right] = 0 \tag{8}$$

and

$$\frac{\partial^2 j_z}{\partial t^2} - A_0^2 \frac{\partial^2 j_z}{\partial z^2} - \nabla^2 \left[(\nu + \nu_e) \frac{\partial j_z}{\partial t} - \nu \nu_e \nabla^2 j_z \right] = 0. \tag{9}$$

Equations for the transverse components are

$$\frac{\partial^2\omega_x}{\partial t^2} - A_0^2 \frac{\partial^2\omega_x}{\partial z^2} - \nabla^2\left[(\nu + \nu_e) \frac{\partial\omega_x}{\partial t} - \nu\nu_e\nabla^2\omega_x\right] = \frac{A_0^2}{2\rho_0} \frac{\partial^3\rho}{\partial t\,\partial y\,\partial z} \tag{10}$$

$$\frac{\partial^2\omega_y}{\partial t^2} - A_0^2 \frac{\partial^2\omega_y}{\partial z^2} - \nabla^2\left[(\nu + \nu_e) \frac{\partial\omega_y}{\partial t} - \nu\nu_e\nabla^2\omega_y\right] = - \frac{A_0^2}{2\rho_0} \frac{\partial^3\rho}{\partial t\,\partial x\,\partial z}, \tag{11}$$

and

$$\frac{\partial^2 j_x}{\partial t^2} - A_0^2 \frac{\partial^2 j_x}{\partial z^2} - \nabla^2\left[(\nu + \nu_e) \frac{\partial j_x}{\partial t} - \nu\nu_e\nabla^2 j_x\right] = \frac{A_0^2}{\mu_e H_0} \frac{\partial^3\rho}{\partial t^2\,\partial y} \tag{12}$$

$$\frac{\partial^2 j_y}{\partial t^2} - A_0^2 \frac{\partial^2 j_y}{\partial z^2} - \nabla^2\left[(\nu + \nu_e) \frac{\partial j_y}{\partial t} - \nu\nu_e\nabla^2 j_y\right] = - \frac{A_0^2}{\mu_e H_0} \frac{\partial^3\rho}{\partial t^2\,\partial x}. \tag{13}$$

These equations show that

(a) the wave motion of vorticity and current density ignore the bulk viscosity λ;

(b) while the transverse components are coupled to the density oscillations and convected with the fluid, the longitudinal components ignore these density oscillations and are diffused as they propagate along lines of force (in both directions). The nature of diffusion due to the combined effects of electrical conductivity and viscosity is very complex due to the higher spatial derivatives introduced by the terms $\nu\,\nu_e\,\nabla^2(\nabla^2)$. It is noticed that equations of the type

$$\nabla^2\left(A\,\frac{\partial\rho}{\partial t} + B\nabla^2\rho\right) = 0 \tag{14}$$

(A and B are constants) have been studied and written about, especially by Oseen [37] but nothing is known about solutions of equations of the type (8).

(c) the symmetrical role played in these equations by the coefficients of kinematic viscosity ν and magnetic viscosity ν_e is remarkable. One can clearly see now why either frictional term can be neglected, depending on either $\nu|\nu_e \ll 1$ or

$\nu|\nu_e >> 1$. On the other hand, if $\nu|\nu_e \cong 1$, but both ν and ν_e are small enough as to neglect their product $\nu\nu_e$, it appears that Eqs. (8)-(13) reduce to Eqs. (9)-(14) of Part III provided that one replaces ν_e with $\nu + \nu_e$.

26. Wave-Motion Equation for the Hydrodynamic Density

The density ρ satisfies

$$\frac{\partial^2 \rho}{\partial t^2} - \nabla^2 \left(a_0^2 + \frac{\lambda + 2\mu}{\rho_0} \frac{\partial}{\partial t} \right) \rho = \mu_e H_0 \left(\frac{\partial j_x}{\partial y} - \frac{\partial j_y}{\partial x} \right). \tag{15}$$

Elimination of j_x, j_y between Eqs. (12), (13), and (15) yields

$$\begin{aligned}
&\frac{\partial^2}{\partial t^2} \left\{ \frac{\partial^2}{\partial t^2} - A_0^2 \frac{\partial^2}{\partial z^2} - \nabla^2 \left[(\nu + \nu_e) \frac{\partial}{\partial t} - \nu\nu_e \nabla^2 \right] \right\} \rho \\
&- a_0^2 \nabla^2 \left\{ \frac{\partial^2}{\partial t^2} - A_0^2 \frac{\partial^2}{\partial z^2} - \nabla^2 \left[(\nu + \nu_e) \frac{\partial}{\partial t} - \nu\nu_e \nabla^2 \right] \right\} \rho \\
&- \frac{\lambda + 2\mu}{\rho_0} \frac{\partial}{\partial t} \left\{ \frac{\partial^2}{\partial t^2} - A_0^2 \frac{\partial^2}{\partial z^2} - \nabla^2 \left[(\nu + \nu_e) \frac{\partial}{\partial t} - \nu\nu_e \nabla^2 \right] \right\} \rho \\
&= A_0^2 \nabla^2 \frac{\partial^2}{\partial t^2} \left(\frac{\partial^2}{\partial x^2} + \frac{\partial^2}{\partial y^2} \right) \rho .
\end{aligned} \tag{16}$$

This equation generalizes Lighthill's [32] equation for $\operatorname{div} \mathbf{v}$.

Using this result, one may now eliminate ρ between Eqs. (10)-(13) and (15), to obtain wave-motion equations for the transverse components of $\boldsymbol{\omega}$ and $\mathbf{j}$.

It is not difficult to write down these equations. Due, however, to their considerable length, and the little information about the structure of their solutions, they are omitted here.

REFERENCES

[1] H. Alfvén, *Arkiv Mat. Astron. Fysik* 29B, No. 2 (1942).
[2] H. Alfvén, "Cosmical Electrodynamics," Chapter 4. Oxford Univ. Press (Clarendon), London and New York, 1950.

[3] P. Appell, "Traité de Mécanique Rationnelle," Vol. III, 2nd ed., pp. 452-455. Gauthier-Villars, Paris, 1909.
[4] A. Baños, Jr., *Phys. Rev.* 97, 1435 (1955).
[5] A. Baños, Jr., *Proc. Roy. Soc.* (London) A233, 350 (1955 and 1956).
[6] A. A. Blank, see ref. [25].
[7] L. Brillouin and M. Parodi, "Propagation des Ondes dans les Milieux Périodiques," see Chapter 6. Masson, Paris, 1956.
[8] J. Carstoiu, *J. Ratl. Mech. Anal.* 3, 691 (1954).
[9] J. Carstoiu, *Proc. Natl. Acad. Sci. U.S.* 46, 131 (1960).
[10] J. Carstoiu, *Compt. Rend.* 251, 509 (1960).
[11] J. Carstoiu, *Proc. Natl. Acad. Sci. U.S.* 47, 891 (1961).
[12] J. Carstoiu, *Compt. Rend.* 252, 2070 (1961).
[13] J. Carstoiu, *Compt. Rend.* 253, 1397 (1961).
[14] J. Carstoiu, *Compt. Rend.* 253, 1653 (1961).
[15] S. Chandrasekhar, "Hydrodynamic and Hydromagnetic Stability," Oxford Univ. Press (Clarendon), London and New York, 1961.
[16] G. Cowling, "Magnetohydrodynamics." Wiley (Interscience), New York, 1957.
[17] G. Cowling, *in* "The Sun" (G. P. Kuiper, ed.), pp. 546-550. Univ. of Chicago Press, Chicago, 1953.
[18] J. W. Dungey, "Cosmic Electrodynamics," Chapter 3. Cambridge Univ. Press, London and New York, 1958.
[19] W. M. Elsasser, *Phys. Rev.* 70, 205 (1946).
[20] W. M. Elsasser, *Phys. Rev.* 72, 828 (1947).
[21] W. M. Elsasser, *Phys. Rev.* 78, 823 (1950).
[22] W. M. Elsasser, *Rev. Mod. Phys.* 28, 135 (1956).
[23] W. M. Elsasser, *in* "Magnetohydrodynamics," First Lockheed Symposium (R. K. M. Landshoff, ed.), see p. 17. Stanford Univ. Press, Stanford, California, 1957.
[24] S. Goldstein, "Lectures on Fluid Mechanics." Wiley (Interscience), New York, 1960.
[25] H. Grad, *in* "Magnetodynamics of Conducting Fluids," Third Lockheed Symposium on Magnetohydrodynamics (D. Bershader, ed.), see pp. 48-58. Stanford Univ. Press, Stanford, California, 1959.
[26] H. Grad and A. A. Blank, "Fluid Dynamical Analogies." NYU, Inst. Math. Sci. Rept., see pp. 48-58 (1958).
[27] H. Grad, "Microscopic and Macroscopic Models in Plasma Physics. NYU, Inst. Math. Sci. Rept. (1961).

[28] N. Herlofson, *Nature* 165, 1020 (1950).
[29] K. U. Ingard, *J. Acoust. Soc. Am.* 31, 1033 (1958).
[30] H. Lamb, "Hydrodynamics," 1st Am. ed. Dover, New York, 1945.
[31] B. Lehnert, *Arkiv Fysik* 5, 63 (1952).
[32] M. J. Lighthill, *Phil. Trans. Roy. Soc. London* A252, 397 (1960).
[33] M. J. Lighthill, *J. Fluid Mech.* 9, 465 (1960).
[34] G. S. S. Ludford, *J. Fluid Mech.* 5, 387 (1959).
[35] S. Lundquist, *Arkiv Fysik* 5, 297 (1952).
[36] G. J. F. MacDonald, *J. Geophys. Res.* 66, 3671 (1961).
[37] C. W. Oseen, *Acta Math.* 34, 205 (1910), see p. 207.
[38] M. Parodi, see ref. [7].
[39] H. Poeverlein, *Phys. Fluids* 4, 397 (1961), see p. 403, Eq. (28).
[40] E. L. Resler, Jr. and J. E. McCune, *Rev. Mod. Phys.* 32, 848 (1960).
[41] E. L. Resler, *Rev. Mod. Phys.* 32, 866 (1960).
[42] J. G. J. Scholte, *J. Atmospheric Terrest. Phys.* 17, 320 (1960).
[43] W. R. Sears, *ARS (Am. Rocket Soc.) J.* 29, 397 (1959).
[44] L. Spitzer, Jr., "Physics of Fully Ionized Gases." Wiley (Interscience), New York, 1956.
[45] T. Stuart, see ref. [30], p. 249.
[46] P. A. Sweet, *Monthly Notices Roy. Astron. Soc.* 110, 69 (1950), see p. 69.
[47] C. Truesdell, *Phys. Rev.* 78, 823 (1950).
[48] C. Truesdell, "The Kinematics of Vorticity." Indiana Univ. Press, Bloomington, Indiana, 1954.
[49] H. C. van de Hulst, "Problems of Cosmical Aerodynamics," Chapter IX. Central Air Documents Office, Dayton, Ohio, 1951.
[50] H. Villat, "Leçons sur les Fluides Visqueux," see Chapter II, pp. 76-77. Gauthier-Villars, Paris, 1943.
[51] C. Walén, *Arkiv Mat. Astron. Fysik* 30A, No. 15 (1944).
[52] H. Weyl, "Space, Time of Matter," pp. 74-75. Dover, New York, 1922.

DEFICIENCIES IN DORODNICYN'S METHOD OF INTEGRAL RELATIONS

E. W. Schwiderski

U. S. Naval Weapons Laboratory*
Dahlgren, Virginia

1. Introduction

In recent years "Dorodnicyn's method of integral relations" [3] has frequently been applied to the so-called detached shock problem (see, e.g. [1, 5, 6, 10]) of aerodynamics. Because of the nonlinearity of this free-boundary value problem of transonic flow past blunt bodies, any numerical method leads to a system of algebraic equations of higher order, for which a well-known theorem prohibits a "direct solution." Thus all attempts at solving this problem must begin the integration with some *a priori* unknown assumption about the flow, for instance, at the detached shock wave [7 - 9, 11] or elsewhere, such that an iteration procedure, which can be applied to the assumed data, converges to the desired solution. Hence any numerical solution of the detached shock problem belongs necessarily to the class of "iterative methods" or, as the aerodynamicists say, to the class of "indirect methods." Nevertheless, Dorodnicyn's method, which has been applied to the detached shock problem by Belotserkovskii and others [1, 5, 6, 10], has always been called a "direct solution" of the shock problem.

*This work was sponsored by the Special Projects Office of the Bureau of Naval Weapons.

It is the intention of the following investigations to resolve this contradiction. Simultaneously it will be shown that the application of Dorodnicyn's method of integral relations to boundary value problems of partial differential equations of mixed elliptic and hyperbolic type is generally in error.

2. Simple Examples of Deficiency

In order to investigate the applicability of Dorodnicyn's method of integral relations to partial differential equations of mixed elliptic and hyperbolic type we may consider the following differential equation of second order:

$$y^{\alpha}\psi_{xx} - \psi_{yy} = 0, \tag{1}$$

or the two equivalent differential equations of first order:

$$u_x + v_y = 0, \qquad y^{\alpha}v_x + u_y = 0, \tag{2}$$

where $u = \psi_y$ and $v = -\psi_x$ and $\alpha = 1$ or -1. In both cases the differential equations (2) are elliptic for $y < 0$ and hyperbolic for $y > 0$. Their "regular" real characteristics are represented by

$$y = \left[\frac{2+\alpha}{2}\,(c \pm x)\right]^{2/(2+\alpha)}, \tag{3}$$

where c is an arbitrary constant.

For $\alpha = 1$ we are dealing with Tricomi's equation which rests on a well-known theory (see, e.g. [2, 4]). For $\alpha = -1$ we are essentially dealing with Dorodnicyn's equation [3], which has not been deeply investigated as yet. Any extrapolation from Tricomi's equation to Dorodnicyn's equation appears most risky, because for $\alpha = -1$ the parabolic line $y = 0$ represents the envelope of the regular characteristics (3) and is thus a "singular" characteristic. Although in both cases fundamental solutions [2, 4] can easily be found in the same manner, it is important to note that these solutions may propagate singularities along characteristics only. Consequently fundamental solutions of (3) may become singular at the entire parabolic line in Dorodnicyn's case but never in Tricomi's case.

Following Dorodnicyn's method of integral relations [3] we find that any pair of functions $[u(x,y),\ v(x,y)]$, which satisfies the differential equations (2), say, in the semi-infinite strip $S = [0 \le x \le 1;\ y \ge y_0 < 0]$, also fulfills the integro-differential equations

$$\begin{aligned} u(1,y) - u(0,y) + \int_0^1 \frac{\partial v(x,y)}{\partial y}\, dx &= 0 \\ y^{\alpha}[v(1,y) - v(0,y)] + \int_0^1 \frac{\partial u(x,y)}{\partial y}\, dx &= 0\,. \end{aligned} \tag{4}$$

Since the important inverse statement has never been proved and evidently cannot be verified, we note the following intolerable deficiency.

Defect 1: Dorodnicyn's method of integral relations permits the construction of "solutions," which are not solutions of the original problem.

By interchanging the differentiation with the integration in Eqs. (4) we arrive at Dorodnicyn's final set of integro-differential equations

$$\begin{aligned} u(1,y) - u(0,y) + \frac{d}{dy}\int_0^1 v(x,y)\, dx &= 0 \\ y^{\alpha}[v(1,y) - v(0,y)] + \frac{d}{dy}\int_0^1 u(x,y)\, dx &= 0\,. \end{aligned} \tag{5}$$

Since this operation is not generally allowed, we note the second deficiency.

Defect 2: Dorodnicyn's method of integral relations permits the exclusion of desired original solutions.

Defects 1 and 2 obviously advise us not to apply the method of integral relations without further deep investigations. The actual consequences of ignoring this urgent warning may be demonstrated in connection with our present example and in the next chapter by means of the detached shock problem.

In order to solve numerically the integro-differential equations (5) Dorodnicyn suggests the application of appropriate quadrature formulas to the evaluation of the remaining integrals [1, 3, 5, 6, 10]. This leads to a set of ordinary differential equations, which must be solved under adequate initial or boundary conditions. By using not only interior spacing ordinates but also exterior supporting lines, one can easily show the possibility to construct "solutions," which are not solutions of the original problem. This demonstrates the first defect of the method.

The consequences of the second defect may be shown by looking at the so-called first approximation, which uses the simple trapezoidal rule. It follows from Eqs. (5) that

$$\begin{aligned} 2(u_1 - u_0) + (v_0' + v_1') &= 0 \\ 2y^{\alpha}(v_1 - v_0) + (u_0' + u_1') &= 0, \end{aligned} \tag{6}$$

where $u_k(y) = u(k,y)$ and $v_k(y) = v(k,y)$ with $k = 0, 1$. The system (6) yields solutions which are regular analytic for all values of y except perhaps for $y = 0$, where the solution may be singular in Dorodnicyn's case but not in Tricomi's case. Consequently the method of integral relations has excluded all solutions which are singular at any regular characteristic. Thus a uniquely defined problem can easily be carried to a wrong solution or to no solution at all, exactly as the defects 1 and 2 do advise us.

As an explicit example consider Dorodnicyn's case $\alpha = -1$ with the special solution

$$u = \frac{-2}{\sqrt{(x-2)^2 - 4y}}, \qquad v = \frac{2 - x}{\sqrt{(x-2)^2 - 4y}}, \tag{7}$$

which is regular analytic at the parabolic line in the integration strip S, but which is singular at the regular characteristic

$$y = \left(\frac{x}{2} - 1\right)^2 . \tag{8}$$

Adding the following boundary values of (7) at $x = 0$ and $x = 1$,

$$v_0 = -u_0 \quad \text{and} \quad u_1 = -2v_1 , \tag{9}$$

to the differential equations (6), then one finds the only solution, which crosses continuously the parabolic line, in the form

$$u_0 - 2v_1 = A(2y)^4 \sum_{m=0}^{\infty} (-1)^m \frac{(2m+8)!}{m!(m+4)!^2} (2y)^m$$

$$v_1 - u_0 = -\frac{A}{6}(2y)^5 \sum_{m=0}^{\infty} (-1)^m \frac{(2m+10)!}{m!(m+5)!^2} (2y)^m, \tag{10}$$

where A stands for one integration constant.

Now if we prescribe v at the lower boundary of our semi-infinite strip S according to the solution (7) by

$$v(x, y_0) = \frac{2 - x}{\sqrt{(x-2)^2 - 4y_0}}, \tag{11}$$

then we obtain no solution at all. If we assume instead of (11) the homogeneous condition

$$(x-2)\, u(x, y_0) = 2v(x, y_0), \tag{12}$$

then all "solutions" are represented by Eqs. (9) and (10) for any value of A, but they all have no resemblance to the correct solution.

3. The Detached Shock Problem

In the detached shock wave problem we are concerned with the nonpurely supersonic flow of an ideal and polytropic gas past a prescribed axisymmetric blunt body of finite dimensions, which generates a certain shock line (see, e.g. [1, 5 - 11]). Although mathematically neither an existence theorem nor an uniqueness proof has been found as yet, it is empirically certain that one and only one solution to this problem exists, provided the disturbance of the body is required to vanish sufficiently fast at infinity. If this important requirement is ignored, our problem will yield a manifold of solutions, exactly as in the analogous subsonic flow theory. This condition causes the disappearance of the shock at least at infinity and generates a singularity along the limiting characteristic, which bounds the dependence of the transonic flow upon the shape of the shock line [8, 9]. In order to solve the detached shock problem by Dorodnicyn's method of integral relations we assume adequate independent variables (x, y) such that $x = 0$ corresponds to the prescribed body contour, $x = 1$ to the *a priori* unknown shock line, and $y = 0$ to the axis of symmetry. If (u, v) denotes certain coordinates of the velocity, then Euler's equations of steady axisymmetric motion can be written in the form [1, 5 - 7, 10]

$$\frac{\partial H_1}{\partial x} + \frac{\partial H_2}{\partial y} = 0, \qquad \frac{\partial G_1}{\partial x} + \frac{\partial G_2}{\partial y} = G_0, \tag{13}$$

where the H_i and G_i denote well-known nonlinear functions of $x, y, u,$ and v.

By integrating Eqs. (13) in the well-known fashion and applying the boundary conditions at the body and at the shock line, we arrive at $2n$ ordinary differential equations for $2n$ dependent functions of y, provided $(n - 1)$ interior spacing lines are used to support the accuracy. Since the axis of symmetry is a streamline, which implies one condition on u and v, only n conditions are available to determine the solution from the axis. The second set of n initial conditions is missing. However, the set of ordinary differential equations becomes singular at a certain line above the axis. Hence Dorodnicyn requires continuity for the solution at this "singular line" and hopes to obtain a second set of boundary conditions, which are sufficient to determine one and only one solution [1, 5, 6, 10].

Before examining the significant deficiencies of this procedure, we first wish to mention a more formal discrepancy. Since it is impossible to solve "directly" the remaining highly nonlinear and singular boundary value problem, one is compelled to assume a complete set of initial flow data at the axis of symmetry, which must be improved by iteration. Thus Dorodnicyn's method of integral relations poses no example of contradiction to mathematics as was pointed out in the introduction; this procedure attacks the detached shock problem in precisely the same "indirect" manner as the solutions described in [7] and [11]. Forced by the extreme difficulties [10], which this general procedure poses for programming on electronic calculators, one is compelled to maintain a fixed number of supporting interior spacing lines (usually none and occasionally one or two). This inflexibility is obviously an extremely risky property as a program of this kind does not permit the well-known numerical error control through change of spacing, which is indispensable in this mathematically undeveloped problem.

The intolerable consequences of defects 1 and 2, which have been overlooked, are as easily found in this example as in the problems of the preceding chapter. Consider first the "singular line" which has been called a "mysterious line" by Van Dyke [12]. This line is a product of the applied method of integral relations and depends completely upon the chosen coordinates of space (x,y) and of velocity (u,v). For instance, if v denotes the total velocity and u some other coordinate, then the "singular line" coincides entirely with the sonic line. Hence the resulting "solutions" are not uniquely defined. This is by no means surprising as the procedure does not employ the boundary condition at infinity, which specifies the flow uniquely. In this connection it is interesting to note that the same phenomenon can be observed in wind tunnel tests. The flow past an obstacle in a wind tunnel definitely depends on the diameter of the tunnel, that is, on the condition at "infinity." Naturally, if a relatively strong shock hits the walls of the tunnel, no singular limiting characteristic can develop which exists under our conditions [8, 9].

Furthermore, it never has been proved that the remaining ordinary boundary value problem possesses one and only one solution, which is by no means trivial. Indeed, consider the same problem in the coordinates used in [7, see Eqs. (14)] and one finds a closed door for any solution. Again this is not surprising, it is very analogous to the solution (7) considered in the previous chapter.

Our results can be summarized as follows:

The Dorodnicyn-Belotserkovskii "solution" of the detached shock problem is indirect and mathematically unfounded. It ignores the determining boundary condition at infinity, but it introduces a continuity condition at a singular line which does not exist in the original problem. It does not show the existence of the singular limiting characteristic and the existence of the point of flow separation behind the shoulder of a body.

REFERENCES

[1] Belotserkovskii, O. M., Flow around a symmetrical profile with detached shock wave, *Prik. Mat. i Mekh.* 22, No. 2 (1958) [English Translation: *Appl. Math. and Mech.* No. 2 (1958)].

[2] Bers, L., "Mathematical Aspects of Subsonic and Transonic Gas Dynamics." Wiley, New York, 1958.

[3] Dorodnicyn, A. A., Method of integral relations for the numerical solution of partial differential equations, Report of the Institute of Exact Mechanics and Computing Technique of the Academy of Sciences of the USSR, Moscow, 1958 (in English).

[4] Guderley, K. G., "Theorie Schallnaher Strömungen." Springer, Berlin, 1957.

[5] Hayes, W. D., and Probstein, R. F., "Hypersonic Flow Theory." Academic Press, New York, 1959.

[6] Holt, M., and Gold, R., Calculation of supersonic flow past a flat-headed cylinder by Belotserkovskii's method, Division of Applied Mathematics, Brown Univ. AFOSR-TN-59-199 (1959).

[7] Schwiderski, E. W., Transonic flows past blunt bodies of revolution with detached shock waves, NWL Rept. No. 1673 (1959).

[8] Schwiderski, E. W., Singularities in supersonic flows past bodies of revolution, NWL Rept. No. 1792 (1962).

[9] Schwiderski, E. W., An asymptotic representation of supersonic flows past bodies of revolution, NWL Rept. No. 1795 (1962).

[10] Traugott, S. C., An approximate solution of the direct supersonic blunt-body problem for arbitrary axissymmetric shapes, *J. Aero/Space Sci.* 27, No. 5 (1960).

[11] Van Dyke, M. D., and Gordon, Helen D., Supersonic flow past a family of blunt axisymmetric bodies, NASA Rept. No. 1 (1959).

[12] Van Dyke, M. D., Review of "Hypersonic Flow Theory," *J. Fluid Mech.* 8, Part 2 (1960).

ON THE STRUCTURE OF HYDRODYNAMIC AND ELECTRODYNAMIC FIELDS

Keith Leon McDonald

University of Utah
Salt Lake City, Utah

1. Introduction

The present work treats the topology of vector fields. A general time-dependent vector function $\mathbf{V}(P,t)$ is analyzed by means of a Taylor expansion written in terms of vector modes, $\mathbf{V}_{(n)}(P,t)$,

$$\mathbf{V} = \hat{\imath}_s V^s(P,t) = \mathbf{V}_{(0)} + \mathbf{V}_{(1)} + \mathbf{V}_{(2)} + \cdots + \mathbf{V}_{(n)} + \cdots . \quad (1)$$

The sth scalar component of the nth mode is a polynomial of degree n in the rectangular coordinates x^1, x^2, x^3, namely,

$$V^s_{i_1 i_2 \ldots i_n} x^{i_1} x^{i_2} \ldots x^{i_n} ;$$

$$V^s_{i_1 i_2 \ldots i_n}(t) = \frac{1}{n!} \left[\frac{\partial^n V^s(P,t)}{\partial x^{i_1} \ldots \partial x^{i_n}} \right]_{P_0} \quad (2)$$

n (not summed) $= 1, 2, \cdots$.

Field structures corresponding to each vector mode are studied and classified for the first several modes of lowest order.

In an earlier analysis (1954) the author discussed the linear fields of a solenoidal vector. The present work was written to complement these fields; the divergence vectors have been included and the analysis is extended to higher order fields.

The properties of vector singular points and nodal points (recalled below) are further described. Regions of considerable importance in vector fields, with respect to physical activity are the neighborhoods of these singularities, other points, in general, being surrounded by constant translation vectors $\mathbf{V}_{(0)}$, in (1). It is therefore essential to analyze the asymptotic field structures in the neighborhoods of vanishing $\mathbf{V}$, for each vector mode. This procedure accounts for missing lower order modes. Of especial interest are the homology structures which form series of fields ranging over successive orders.

Concerning previous investigations, the misconceptions which frequently parallel the concept of assemblages of lines of force of divergenceless vectors was pointed out by Slepian (1951). Elsasser and McDonald (1953) briefly discussed the applications of related concepts to magnetohydrodynamics, and in a later paper (1954) the author treated the topology of linear, stationary current, magnetic fields. This work has been partly complemented in a series of four technical reports (McDonald, 1959; 1960, a, b; 1962) by extending the analysis to higher order modes for both the solenoidal and divergence cases, and treating physical applications. The aim of this paper is of more limited scope, it being an abridgement which attempts to establish or summarize structural traits rather than to emphasize physical analyses. Hence, the reader will frequently be referred to the above references for detailed derivations and applications. In Section 2, we have omitted numerous exact polynomial solutions of the hydrodynamic equations and included only the first-, second-, and third-order deformation fields to illustrate a typical homology series. Numerous other possible asymptotic fields which are not solutions of any particular set of differential equations have likewise been referred to the more extensive work. The confluent vector singular point is introduced in connection with the Hertzian dipole oscillator, which example also establishes that motions of singularities in the dynamic field are not limited by the speed of light. An account of holomorphic functions completes this section and the non-occurrence of nodal points is briefly qualified in Section 3.

We conclude this section by recalling the properties of the two well-known principal types of singular points (McDonald, 1954), as follows: (a) A vector field $\mathbf{V}$ contains *vector singular points* $\{P_0\}$ provided that $\mathbf{V}$ is bounded and the direction of $\mathbf{V}$ is not unique. In physical applications, V must vanish at P_0. The only remaining nontrivial singularity is (b) the *infinite singular point* (V unbounded). Examples of trivial singular points are found at the boundary separating two media of different material parameters. Thus, a jump in the magnetic permeability gives rise to a "kink" in the lines of magnetic intensity. In a regular domain of $f(z)$ in the complex plane, a point z_0 is a vector singularity if, and only if, $f'(z_0) = 0$. Nodal points cannot occur. A branch point z_0 is defined as a vector singularity if limit $f'(z) \to 0$, as $z \to z_0$, and the branch cut is regarded as a thin strip whose edges delimit the boundary of a physical domain on its two sides. The functions $z^{i\alpha}$, $z^{\alpha z}$, α = constant, establish that bounded V is either oscillatory divergent or it approaches a nonvanishing limit for branch points $\{z_0\}$ which are not vector singularities.

2. Field Structures

Individual vector modes in (1) are effectively written in matrix form. Thus, choosing principal axes of the deformation vector, one writes the linear fields as

$$\begin{bmatrix} V^1_{(1)} \\ V^2_{(1)} \\ V^3_{(1)} \end{bmatrix} = \begin{bmatrix} (a+b) & -\Omega_0^{\,3} & \Omega_0^{\,2} \\ \Omega_0^{\,3} & (a-b) & -\Omega_0^{\,1} \\ -\Omega_0^{\,2} & \Omega_0^{\,1} & c \end{bmatrix} \cdot \begin{bmatrix} x^1 \\ x^2 \\ x^3 \end{bmatrix} \equiv \mathsf{A} \cdot \mathsf{X}, \qquad (3)$$

the rotation vector being described by the three antisymmetric elements. When $\mathbf{V}_{(1)}$ is solenoidal, the trace of the coefficient matrix A vanishes $(c = -2a)$ and the resulting parameters specify five independent component fields. Their classification has already been discussed (McDonald, 1954). Although not a unique representation, they are sufficient to describe any solenoidal linear field. In similar manner the second-order coefficient matrix, typifying the form of the higher order matrices, comprises 18 terms,

$$\mathsf{B} = \begin{bmatrix} B^1_{11} & B^1_{12} & B^1_{22} & B^1_{23} & B^1_{13} & B^1_{33} \\ B^2_{11} & B^2_{12} & B^2_{22} & B^2_{23} & B^2_{13} & B^2_{33} \\ B^3_{11} & B^3_{12} & B^3_{22} & B^3_{23} & B^3_{13} & B^3_{33} \end{bmatrix} . \tag{4}$$

The coefficient relations are, by reason of the indifference to order of differentiation in the Taylor coefficients,

$$B^i_{jk} = \frac{2!}{1!} V^i_{jk}, \quad j \neq k; \quad B^i_{jj} = \frac{2!}{2!} V^i_{jj} \quad (j \text{ not summed}). \tag{5}$$

A reduction from 9 to 6 independent parameters in the first-order coefficient matrix obtains by a transformation to principal axes. Likewise, the matrices of the second- and third-order fields reduce from 18 to 15 coefficients and from 30 to 27, respectively, whenever symmetry axes are chosen. In general, however, the *principal axes* of the linear modes do not coincide with any of the several sets of *symmetry axes* of the higher order modes. Since intrinsic properties only are desired, symmetry axes are chosen in each case.

The eigenvalue problem which requires the diagonalization of the characteristic matrix is well known in the mechanics of continuous media and crystal optics. Generalizing this scheme, employed in the linear fields, one may locate the sets of symmetry axes by forming the scalar dot product of $\mathbf{V}_{(n)}$ with the position vector $\mathbf{r}$. This is illustrated here for the second-order mode (4):

$$\begin{aligned} \mathbf{r} \cdot \mathbf{V}_{(2)} = {} & x^1\left[B^1_{11}(x^1)^2 + \left(B^1_{12} + B^2_{11}\right)x^1x^2 + \left(B^2_{12} + B^1_{22}\right)(x^2)^2\right] \\ & + x^2\left[B^2_{22}(x^2)^2 + \left(B^2_{23} + B^3_{22}\right)x^2x^3 + \left(B^3_{23} + B^2_{33}\right)(x^3)^2\right] \\ & + x^3\left[B^3_{33}(x^3)^2 + \left(B^3_{31} + B^1_{33}\right)x^1x^3 + \left(B^1_{31} + B^3_{11}\right)(x^1)^2\right] \\ & + x^1x^2x^3\left[B^3_{12} + B^2_{13} + B^1_{23}\right] . \end{aligned} \tag{6}$$

It is assumed that (6) has been transformed from a y-coordinate system according to the orthogonal transformation

$$x^i = a^i_j\, y^j, \qquad y^j = \alpha^j_i\, x^i; \qquad \alpha^j_i = a^i_j. \tag{7}$$

Among the nine direction cosines a^i_j there exist six independent relations, thereby leaving three arbitrary choices among the B^k_{ij}. For example, one may choose to nullify the coefficients of the three mixed terms within the brackets. The resulting symmetry axes are specified by

$$B^1_{12} = -B^2_{11}, \qquad B^2_{23} = -B^3_{22}, \qquad B^3_{31} = -B^1_{33}. \tag{8}$$

Other symmetry choices are obvious by inspection of (6). The remaining coefficients may be further restricted by imposing the divergenceless condition*

$$\nabla \cdot \mathbf{V}_{(n)} = 0, \quad \text{or} \quad \mathbf{V}^{\alpha}_{\alpha i_2 i_3 \cdots i_n} = 0. \tag{9}$$

By imposing (8) and (9) the matrix (4) simplifies from 18 to 12 independent coefficients, which represents, therefore, 12 independent solenoidal vector fields of second order.

An extensive analysis of the lower order modes has established numerous field structures, including several exact polynomial solutions of the full hydrodynamic equations. These are elucidated in the more extensive analyses (see references) by means of assemblages of vector lines of force. We indicate

---

*Our analysis (McDonald, 1960) of the equation of continuity, $\partial\rho/\partial t + \nabla \cdot (\rho\mathbf{v}) = 0$, for a steady state compressible fluid, shows that the velocity mode of lowest order present in the vector Taylor expansion, the asymptotic field, is divergenceless. Hence (9) is particularly appropriate. Moreover, a fact which contributes measurably to the utility of vector mode is the following:

$$\nabla \cdot \mathbf{V} = 0, \quad \text{implies that} \quad \nabla \cdot \mathbf{V}_{(n)} = 0, \quad n = 0, 1, 2, \cdots.$$

Each vector mode is individually solenoidal if $\mathbf{V}$ *is solenoidal, and conversely.* This result is established by recognizing that each mode of order n consists of three scalar component homogeneous polynomials in x^1, x^2, x^3 of degree n.

here only the first three fields of an important homology series, the deformation fields shown in Figs. 1 - 3. Figure 1 depicts the stationary magnetic field of two cylindrical conductors of equal electric currents, I, joined to form a cross. The lines $x^2 = 0$, $x^1 = x^3$ (surrounded by linear deformation fields)

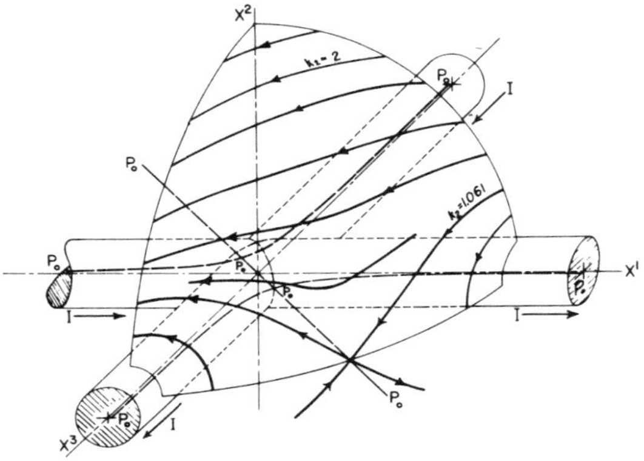

Fig. 1. Stationary current magnetic field of crossed straight-wire conductors, illustrating the linear deformation field.

and $x^1x^3 = I/\pi J$ are lines of vector singular points intersecting at points $(\pm a, 0, \pm a)$, where $a = (I/\pi J)^{1/2}$ is the radius of the conductors. The (closed) lines of force external to the current region lie in the spherical surface, shown in the positive octant. Figures 2 and 3 are exact polynomial solutions of the hydrodynamic equations for a homogeneous fluid. Observe that the flow lines (dashed) intercept the cylindrical isopressure surfaces at arbitrary angles, but intercept the velocity potential surfaces orthogonally. One may readily demonstrate that the differential pressure varies as r^{2n}, n designating the order of the structure in the homology series.

Next let us consider an example taken from electrodynamics. Figure 4 shows a rough sketch of the E-field structure of a sinusoidal Hertz oscillator. Our analysis shows that the singular points form circular line distributions in the equatorial plane, with polar radius prescribed by the transcendental equation

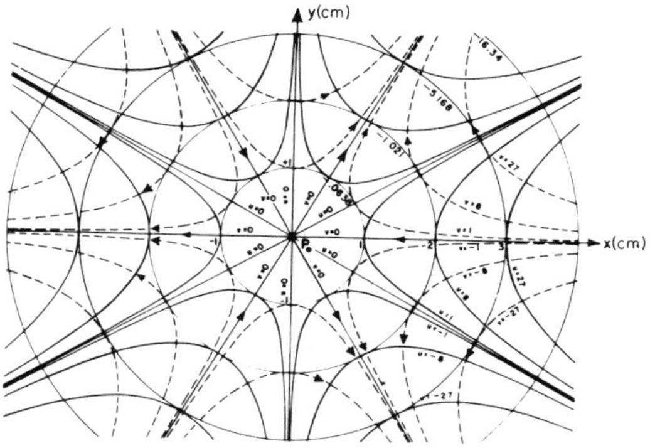

Fig. 2. Second-order deformation field, a two-dimensional mode indifferent to fluid viscosity. $\Delta p = r^4 \times$ constant (stand. atm. $\times\ 10^{-6}$) corresponds to $V^1_{11} = -10\ \mathrm{cm}^{-1}\ \mathrm{sec}^{-1}$; $f(z) = \alpha z^3/3$. Regarding the Bessel function of the first kind, $z^3 \simeq J_3(z) \cdot 2^3 3!$

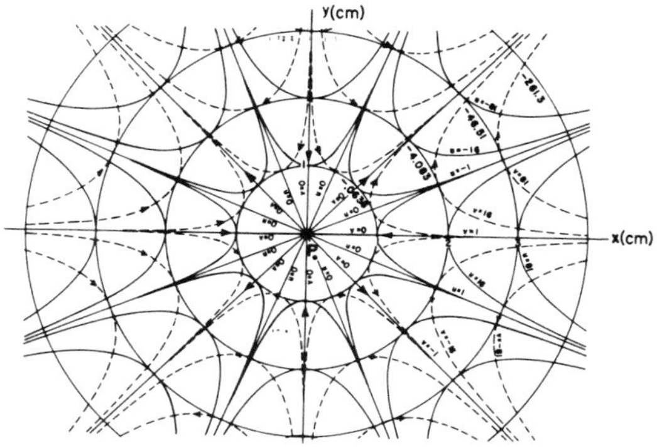

Fig. 3. Third-order deformation field, a mode indifferent to fluid viscosity. $\Delta p = r^6 \times$ constant (stand. atm. $\times\ 10^{-6}$) corresponds to

$$|\mathbf{V}_{(3)}|_{r=1\ \mathrm{cm}} = 10\ \mathrm{cm\ sec}^{-1};$$

$$f'(z) = \alpha z^3;\quad z^4 \simeq J_4(z) \cdot 2^4 4!$$

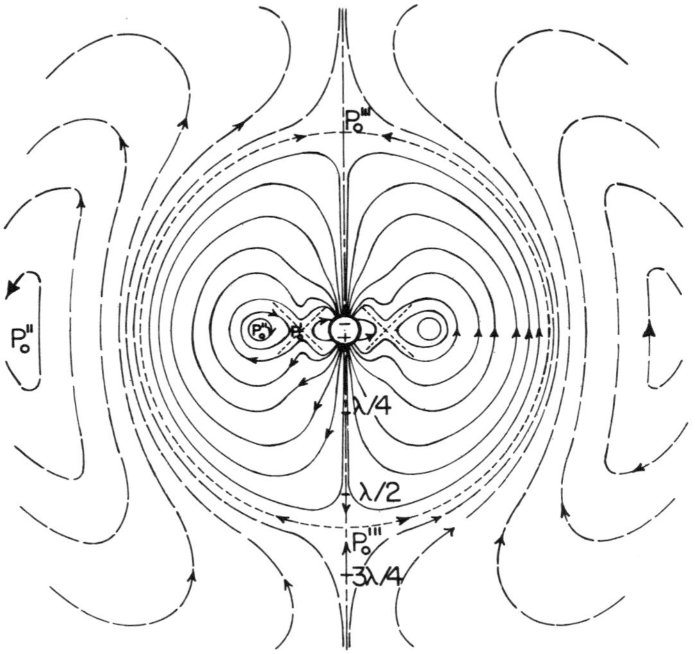

Fig. 4. Rough sketch of meridional view of **E**-field structure of sinusoidal Hertz oscillator (exaggerated central sphere) at time slightly greater than -0.123*T*. [After C. Schaefer, *Einführung in die theoretische Physik*, Band 3, 342 (1932).]

$$Y_1 = Y_2, \quad Y_1(\xi, t) = \tan(\xi - \omega t), \quad Y_2(\xi, t) = \frac{\xi}{1 - \xi^2}; \tag{10}$$

$\xi(\equiv kr)$ is expressed in terms of the circular wave number $k = 2\pi/\lambda$. The entire solution comprises a countable infinity of solutions only one of which appears, during any semicycle, as a ring of confluent vector singular points at $(\xi, t) \doteq (1.43, -0.123T)$ (cf. Fig. 5). At the time of generation, the latter ring separates into two rings, one contracting toward the oscillator (comprising a circular line distribution of linear deformation fields) and one expanding (linear rotation fields), the respective radial velocities at the time of generation being $dr/dt = \mp\infty$, computed from (10). The velocities of the rotation field singularities approach the phase velocity ω/k as $r \to \infty$ from superior

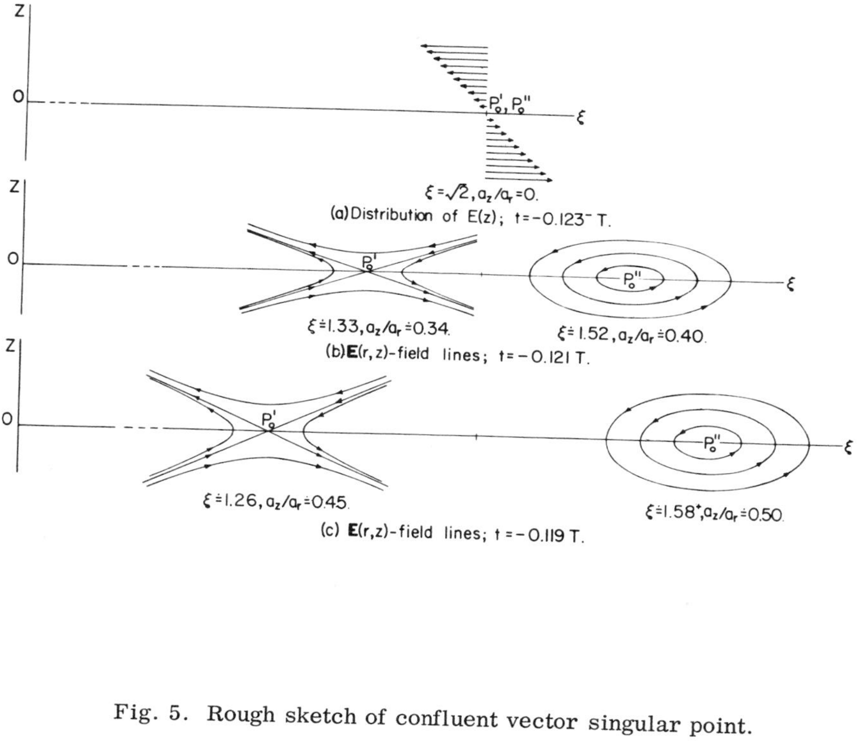

Fig. 5. Rough sketch of confluent vector singular point.

values. The lifetime of the single contracting ring is only a small fraction ($\doteq 0.123T$) of the oscillator period T. The magnitude of the radial velocity first decreases as the oscillator is approached and then again increases to an infinite velocity at time $t = 0-$, where it disappears at the oscillator. This example establishes that *vector singularities may be generated instantaneously in the dynamic field and their motions are not limited by the speed of light*. We conclude this section with a brief account of holomorphic functions.

At regular points of $f(z)$, $z = x + iy$, vector singularities are characterized by the lack of uniqueness of the tangents to the conjugate u-, v-curves, such that non-orthogonal or multiple intersections occur. This anomalous behavior is consistent with the orthogonality condition $\nabla u \cdot \nabla v = 0$, only if $|\nabla u|_0 = 0$ [$|\nabla u|_0 = |\nabla v|_0 = |f'(z_0)|$ by the Cauchy-Riemann relations]. Consequently, the existence of this singular point requires that

$$f'(z_0) = 0. \tag{11}$$

This necessary condition becomes familiar on physical grounds when one recalls that the scalar components of **V** are

$$V_1 = -\partial u/\partial x, \qquad V_2 = -\partial u/\partial y\,; \tag{12}$$

V is parallel to the curves defined by $v(x, y)$ = constant.

Whenever $f(z)$ is regular at z_0, (11) is also a sufficient condition. Excepting the trivial case $f(z)$ = constant [or $f'(z) \equiv 0$] the problem reduces to demonstrating the nonexistence of *nodal points* $Q_0 = z_0$. Now the regularity of $f(z)$ at z_0 ensures a nonvanishing radius of convergence of its Taylor expansion

$$f(z) - a_0 = 0 \cdot (z - z_0)^1 + a_2(z - z_0)^2 + \cdots + a_n(z - z_0)^n + \cdots. \tag{13}$$

Thus, if a_k $(k \geq 2)$ is the first nonvanishing coefficient on the right, then in the neighborhood of z_0, $f(z)$ is asymptotic to $a_0 + a_k(z - z_0)^k$, as $z \to z_0$. Writing (13) in polar notation it becomes evident that throughout the range of values $0 \leq \theta < 2\pi$ there are k identical configurations of u-, v-curves. These are separated by $2k$ straight lines intersecting at $P_0(x_0, y_0)$, for each of the relations $u - |a_0| \cos(\arg a_0) = 0$, $v - |a_0| \sin(\arg a_0) = 0$ (cf. Figs. 2 and 3). We have thus established that *a nodal point* Q_0 *cannot exist in a domain where* $f(z)$ *is regular, and (11) is the n.a.s.c. that* z_0 *be a vector singular point.*

One further consequence is the following: *A vector singular point in the regular domain is isolated and surrounded by a deformation field* (of order $n \geq 1$). This result follows immediately from a well-known theorem on regular functions (Copson, 1935), which states that the n.a.s.c. that a function $f'(z)$ vanish at all points of the domain of regularity is that there exist a sequence of zero points $z_1, z_2, \cdots, z_n, \cdots$, of $f'(z)$ having a limit point interior to D. Evidently, one must exclude from the regular domains the rotation and divergence fields of all orders; the common focus is not allowed; of the three linear two-dimensional fields (McDonald, 1954), only the deformation field is possible, by (13).

There remains to consider branch points, z_0, where $f'(z)$ is not defined; one computes the limiting value as $z \to z_0$ at a sequence of points which has z_0 as limit point:

$$\lim_{z \to z_0} f'(z) = \lambda(z_0). \tag{14}$$

The previous result pertaining to regular functions is now generalized to include branch points by replacing $f'(z_0)$ by $\lambda(z_0)$, as follows: *If a vector singularity exists at a branch point* z_0, *then it is a n.a.s.c. that* $\lambda(z_0) = 0$. For $\lambda(z_0)$ must vanish at a vector singularity in all physical applications, whereas the converse follows from the fact that the neighborhood of a nodal point cannot be asymptotic to that of a branch point because of the essential singular points comprising the branch cut.

3. Nodal Points

General conditions under which nodal points cannot occur are next established. First consideration is given to the divergenceless vector field **V**. The proof, by *reductio ad absurdum*, assumes the existence of an elemental flux tube (whose lateral confines are defined by the vector lines of force) which passes continuously through the void region of an isolated nodal point (Fig. 6). Such a tube would be in violation

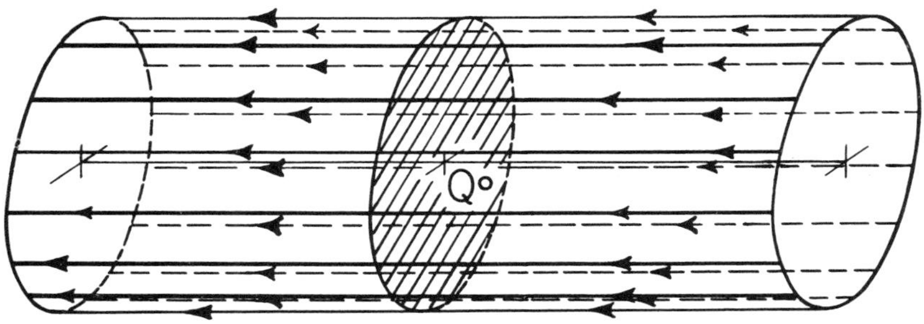

Fig. 6. On the non-occurrence of nodal points.

of the divergence theorem. For one could choose one end face to pass through the nodal point thus resulting in a nonvanishing net flux. This contradiction establishes the result. It follows by the same argument that nodal distributions forming (non-closed) line segments and two-sided (open) surfaces of bounded extent cannot occur. Likewise, limit cycles cannot comprise nodal points. Finally, application of the divergence theorem readily establishes that closed, regular surface, nodal distributions demand by continuity arguments that **V** vanish at all points of the domain of definition (whether singly or multiply connected).* Let us turn to applications.

The non-occurrence of nodal points in electrodynamics is immediately applicable to the magnetic induction **B**, since $\nabla \cdot \mathbf{B} = 0$. A similar argument holds for the electric displacement **D** whenever the free charge density ρ_F vanishes; $\nabla \cdot \mathbf{D} = 0$. Similar conclusions apply to the fields of **H** and **E** by application of the material equations $\mathbf{B} = \mu\mathbf{H}$, $\mathbf{D} = K\mathbf{E}$, valid for conducting and nonconducting homogeneous and isotropic media. This disposes of those important problems in the propagation of electromagnetic waves.

In electrically conducting inhomogeneous media both dynamic and quasi-stationary fields require a nonvanishing charge distribution, as in the hypothetical case of static distributions of free charge. For such cases, conditions under which nodal points may exist are established by two simultaneous applications of the divergence theorem to Fig. 6, one to the left-hand section of the elemental tube, and one to the right-hand section. When one relates the field to its source, the right-hand assemblage of lines infers a net negative charge $Q_- = \int \rho_F \, d\tau$, with a simultaneous positive charge Q_+ in the left-hand section. An electrostatic example which establishes the existence of a planar distribution of nodal points is described in the more extensive work (McDonald, 1962).

Next consider the phenomena of hydrodynamics. In the case of an incompressible fluid $(d\rho/dt = 0)$, the equation of continuity requires that $\nabla \cdot \mathbf{v} = 0$, so that the velocity field is divergenceless and nodal points cannot occur. Similarly, the existence of a nodal point in a steady compressible fluid violates the classical mass conservation principle. For the mass

*The singular cases in which the vector lines of force are parallel to the nodal distributions and their closed line boundary points are not treated here.

flow, $\rho \mathbf{v}\, dS\, dt$, could not be conserved over different cross-sectional areas dS of an elemental flow tube.

Conditions sufficient for the existence of a discrete nodal point in an accelerated compressible fluid are obtained by computing the flux of $\rho \mathbf{v}$ through the two halves of the elemental flow tube; an influx of mass in the right-hand section must accompany a decrease of mass in the left-hand section, an unlikely spontaneous configuration.

The foregoing arguments apply to those velocity fields in elastic solids. Likewise, the theory of elasticity is generally applicable when one replaces the velocity $\mathbf{v}$ by the corresponding displacement vector $\mathbf{d}$. Further qualifications appear unnecessary.

REFERENCES

Copson, E. T. (1935). "Theory of Functions of a Complex Variable." Oxford Univ. Press, London.

Elsasser, W. M., and K. McDonald (1953). Topology of magnetic fields. *Bull. Am. Phys. Soc.* 28, No. 5.

McDonald, K. L. (1954). Topology of steady current magnetic fields. *Am. J. Phys.* 22, 586-596.

McDonald, K. L. (1959). On the structure of hydrodynamic and electrodynamic fields. Part I: Vector expansions in rectangular coordinates. U.S. Army Chemical Corps Proving Ground, TR (non.) DPGR 240, Dugway, Utah.

McDonald, K. L. (1960a). On the structure of hydrodynamic and electrodynamic fields. Part II: Distribution of type (b) singularities in the Euclidean complex plane. U.S. Army Chemical Corps Proving Ground, TR (non.) DPGR 240, Dugway, Utah.

McDonald, K. L. (1960b). Taylor expansions of the hydrodynamic equations, Part I. Department of Physics, Brigham Young University, TR (non.), Provo, Utah.

McDonald, K. L. (1962). Taylor expansions of the hydrodynamic equations, Part II. Department of Physics, University of Utah, TR (non.), Salt Lake City, Utah.

Slepian, J. (1951). Lines of force in electric and magnetic fields. *Am. J. Phys.* 19, 87.

DISTRIBUTION FUNCTIONS FOR MOMENTUM TRANSFER IN AN IDEALIZED PLASMA

T. Triffet

Michigan State University
East Lansing, Michigan

Solutions to a large number of contemporary engineering problems, ranging from applications of superconductivity on the one hand to power studies of fusion reactions on the other, depend on establishing the transfer properties of matter in specific plasma states. In principle, however, all such properties can readily be calculated once appropriate distribution functions

$$f(\mathbf{r},\mathbf{v},t), \tag{1}$$

where

$$f\,d\mathbf{r}\,d\mathbf{v} \tag{2}$$

defines the number of particle points in the range $d\mathbf{r}$ about $\mathbf{r}$ and $d\mathbf{v}$ about $\mathbf{v}$ in six-dimensional $\mathbf{x}$, $\mathbf{v}$ space, have been determined for the separate particle species.

For example, in an idealized electrically neutral plasma consisting only of electrons and singly positive ions of mass m, the ion density will be

$$n = \int f^{+} \, d\mathbf{v} \tag{3}$$

and their mean velocity

$$\mathbf{V} = \frac{1}{n} \int \mathbf{v} \, f^{+} \, d\mathbf{v}. \tag{4}$$

Consequently the momentum vector will be given by

$$P_i = \int m v_i f^{+} \, d\mathbf{v} = nmV_i \tag{5}$$

and the momentum transfer tensor by

$$P_{ij} = \int m v_i v_j f^{+} \, d\mathbf{v} = nm\overline{v_i v_j} , \tag{6}$$

the bar indicating an average over the dyadic product; it follows of course that the pressure tensor, arising from momentum transfer associated with the random particle velocity $(\mathbf{v} - \mathbf{V})$, may be written

$$p_{ij} = \int m(v_i - V_i)(v_j - V_j) f^{+} \, d\mathbf{v} . \tag{7}$$

The pressure tensor for the electrons can be expressed in the same way in terms of $f^{-}(\mathbf{r},\mathbf{v},t)$, analogous to $f^{+}(\mathbf{r},\mathbf{v},t)$; and these functions also define mass and heat flux vectors.

If f is Maxwellian, $\mathbf{V}$ vanishes and Eqs. (6) and (7), for example, lead to [1]

$$p_{ij} = P_{ij} = nkT \, \delta_{ij} , \tag{8}$$

where k is Boltzmann's constant, T the absolute temperature and δ_{ij} the unit tensor; but though this constitutes a useful reference state, nonequilibrium states are necessarily more important in the analysis of transfer properties. It is essential that f be determined as a function of the time t, as well as the position and velocity coordinates $\mathbf{r}$ and $\mathbf{v}$, in most cases of practical interest.

In engineering applications the usual way of attacking this problem has been to attempt a simultaneous solution of Boltzmann's equation, involving the distribution function $f(\mathbf{r}, \mathbf{v}, t)$ of a particular species,

$$\frac{\partial f}{\partial t} + \mathbf{v} \cdot \frac{\partial f}{\partial \mathbf{r}} + \frac{\mathbf{F}}{m} \cdot \frac{\partial f}{\partial \mathbf{v}} = \left[\frac{df}{dt}\right]_{\text{collisions}}, \tag{9}$$

and Maxwell's electromagnetic field equations,

$$\nabla \times \mathbf{H} = \mathbf{J} + \frac{\partial \mathbf{D}}{\partial t}, \qquad \mathbf{D} = \epsilon \mathbf{E} \tag{10}$$

$$\nabla \times \mathbf{E} = -\frac{\partial \mathbf{B}}{\partial t}, \qquad \mathbf{B} = \mu \mathbf{H}, \tag{11}$$

wherein all terms have their usual meanings and the electrical current density $\mathbf{J}$ may also be expressed as

$$\mathbf{J} = \sum_{\text{species}} q \int \mathbf{v} f \, d\mathbf{v}, \tag{12}$$

where q equals the electrical charge of a particle of the species described by f. Clearly the matter present is coupled with the electromagnetic field through this term, while the external force vector $\mathbf{F}$ in Eq. (9) must include both gravitational and electromagnetic contributions:

$$\mathbf{F} = m\mathbf{G} + (q\mathbf{E} + \mathbf{J} \times \mathbf{B}). \tag{13}$$

Two-body interactions are then assumed and the right-hand side of Eq. (9) taken to represent the sum of close and distant collision terms. The first of these may be shown to have the form [2]

$$\left[\frac{df}{dt}\right]_{\substack{\text{close}\\ \text{collisions}}} = 2\pi \sum_{\text{*species}} \iint (f' f'^* - f f^*) \nu b \, db \, d\mathbf{v}^*, \tag{14}$$

where the prime refers to post-collision f values, ν means the initial absolute relative velocity of the two particles under consideration, and b is their distance of closest approach for no interaction. The second, however, is more troublesome, involving the momentum transfer of many small-angle scattering events. In uncharged fluids, especially dilute gases, it can usually be ignored; but not in plasmas, since it may equal or exceed the first term in magnitude. The Fokker-Planck assumptions, operating to bound such processes by statistical considerations, are most often invoked; these lead to a term of the form [3]

$$\left[\frac{df}{dt}\right]_{\substack{\text{distant}\\ \text{collisions}}} = \sum_{*\text{species}} \frac{1}{2}\left[\sum_{i,j} \frac{\partial^2}{\partial v_i \partial v_j} (f <\Delta v_i \Delta v_j>^*) - \sum_{i} \frac{2\partial}{\partial v_i} (f <\Delta v_i>^*)\right], \quad (15)$$

$$<\Delta v_i>^* = 2\pi \int_0^\infty \nu f^* \, dv_i^* \int_{b_{\min}}^{b_{\max}} \Delta v_i b \, db , \quad (16)$$

where $<\Delta v_i>^*$ indicates the change in the ith component of the velocity of a given particle due to scattering by any other in the range established by $b_{\max}$ and $b_{\min}$, which must be determined from physical considerations.

In practice, of course, very considerable difficulties arise in attempting to solve the integro-differential equations which follow from Eqs. (9)-(14), particularly for dense fluids; and when the distant collisions term, Eqs. (15) and (16), is included these are increased many times over [4]. Ultimately so many simplifying assumptions must be made that it only can be concluded the method is not well suited to deal with long-range interactions, and thus can never be entirely satisfactory for plasmas. Accordingly, the principal purpose of this chapter is to emphasize that the physical problem can be formulated in such a way as to pose a different, perhaps more tractable, mathematical problem.

It has been shown by Born and Green [5], Irving and Kirkwood [6], and others that the transfer properties of uncharged but dense fluids can be defined directly from Liouville's equation describing the time evolution of the Nth order distribution function

$$f_N(\mathbf{r}, \mathbf{p}, t) \quad (17)$$

for the representative points in the six-dimensional $\mathbf{x}$, $\mathbf{p}$-space of a system consisting of N particles,

$$\frac{\partial f_N}{\partial t} = -\frac{\mathbf{p}_N}{m} \cdot \frac{\partial f_N}{\partial \mathbf{r}_N} + \frac{\partial \Phi}{\partial \mathbf{r}_N} \cdot \frac{\partial f_N}{\partial \mathbf{p}_N}, \tag{18}$$

where the total potential energy

$$\Phi = \frac{1}{2} \sum_{\alpha=1} \sum_{\beta=1} \phi(r_{\alpha\beta}), \tag{19}$$

by means of δ-function delineation.

From Eq. (18) and the fact that the average value of any dynamic variable $\gamma(\mathbf{r}, \mathbf{p}, t)$ of the system must be

$$\bar{\gamma} = \frac{1}{N!} \iint \gamma f_N \, d\mathbf{r}_N \, d\mathbf{p}_N \tag{20}$$

it follows that

$$\frac{d\bar{\gamma}}{dt} = \overline{\frac{\mathbf{p}_N}{m} \cdot \frac{\partial \gamma}{\partial \mathbf{r}_N} - \frac{\partial \Phi}{\partial \mathbf{r}_N} \cdot \frac{\partial \gamma}{\partial \mathbf{p}_N}}. \tag{21}$$

This general equation of change actually contains Boltzmann's equation, featuring the lower order distribution function $f(\mathbf{r}, \mathbf{v}, t)$, as a special case. Nevertheless, if γ is specified as

$$\gamma = \sum_{\alpha=1}^{N} \mathbf{p}_\alpha \, \delta(\mathbf{r}_\alpha - \mathbf{r}) \tag{22}$$

and the lower order distribution functions are defined in a corresponding manner, it can be demonstrated [6] that Eq. (21) yields the usual equation of motion for fluids,

$$\frac{\partial \mathbf{V}}{\partial t} + \mathbf{V} \cdot \frac{\partial \mathbf{V}}{\partial \mathbf{r}} = \frac{1}{\rho} \left[n\mathbf{F} - \frac{\partial}{\partial \mathbf{r}} : \mathbb{P} \right], \tag{23}$$

providing the pressure tensor $\mathbb{P}$ is given by

$$\mathbb{P} = m \int \left(\frac{\mathbf{p}}{m} - \mathbf{V} \right) \left(\frac{\mathbf{p}}{m} - \mathbf{V} \right) f(\mathbf{r}, \mathbf{p}, t) \, d\mathbf{p}$$

$$- \frac{n^2}{2} \int \frac{d\phi(R)}{dR} \, g(\mathbf{r}, \mathbf{R}, t) \, \frac{\mathbf{R}\mathbf{R}}{R} \, d\mathbf{R}. \tag{24}$$

Here $\mathbf{R}$ is the separation vector of the particle pair under consideration and $g(\mathbf{r},\mathbf{R},t)$ is a radial distribution function for pairs, rather than singles like the velocity and momentum distribution functions $f(\mathbf{r},\mathbf{v},t)$ and $f(\mathbf{r},\mathbf{p},t)$. Corresponding relations for the mass and heat flux vectors can also be written in terms of these same functions [7]. It should be emphasized that a parallel formalism featuring the probability density matrix ρ_{mn}, the quantum mechanical analog of a classical distribution function [8], can be developed for use when quantum effects are important.

As before, $\mathbf{F}$ must be given by Eq. (13),

$$\mathbf{F} = m\mathbf{G} + (q\mathbf{E} + \mathbf{J} \times \mathbf{B}) ,$$

where $\mathbf{E}$ and $\mathbf{J}$ satisfy Maxwell's equations (10) and (11) and the latter couples the matter and electromagnetic field through Eq. (12),

$$\mathbf{J} = \sum_{\text{species}} q \int \mathbf{v} f \, d\mathbf{v} .$$

But the practical advantage of this formulation is that the kinetic and potential contributions to the transfer tensors appear naturally and yet remain distinct; the first term on the right in Eq. (24) represents the former, for example, while the second term represents the latter. This means that the relative importance of each may conveniently be assessed for different fluid states.

It is known, for example, that the first term dominates for dilute gases and the second, which by its nature is weighted in favor of momentum transfer through distant collisions, for dense gases and liquids [7]. Both may be important in plasmas, of course; but by utilizing the concept of an average potential of force on each particle due to all others separately, the Liouville equation (18) may be rewritten in terms of a distribution function of arbitrary order h:

$$\frac{\partial f_h}{\partial t} + \sum_{\ell=1}^{N} \left[\frac{\mathbf{p}_\ell}{m} \cdot \frac{\partial f_h}{\partial \mathbf{r}_\ell} - \frac{\partial \Phi}{\partial \mathbf{r}_\ell} \cdot \frac{\partial f_h}{\partial \mathbf{p}_\ell} \right]$$

$$= \sum_{\ell=1}^{h} \iint \left(\frac{\partial \phi}{\partial \mathbf{r}_\ell} \cdot \frac{\partial f_{h+1}}{\partial \mathbf{p}_k} \right) d\mathbf{r}_{h+1} \, d\mathbf{p}_{h+1} . \qquad (25)$$

As suggested in Eq. (3), the distribution function which appears in the first term of Eq. (24) is related to the number density $n = n(\mathbf{r}, t)$ by

$$n_1 = \int f_1 \, d\mathbf{p} \,, \tag{26}$$

where the order of both has been indicated. Similarly,

$$n_2 = \int f_2 d\mathbf{p} \, d\mathbf{p}^*; \tag{27}$$

and $g(\mathbf{r}, \mathbf{R}, t)$, which appears in the second term of Eq. (24), is related to this by

$$n_2/n_1^* n_1 = g(\mathbf{r}, \mathbf{R}, T). \tag{28}$$

Equation (25) defines f_1 in terms of f_2, f_2 in terms of f_3, and so on; thus if f_3 can be estimated, two simultaneous equations for f_1 and f_2 will result.

More generally, the earlier mathematical problem has been replaced by one of terminating a set of coupled integral equations in some sensible way. Procedures have been devised for dense gases and liquids by Green [9] and Kirkwood [10], and notable work on a method for plasmas has been reported by Rostoker and Rosenbluth [11]. A promising approach featuring a new solution of the Bogoliubov functional form of the Liouville equation has also been suggested recently by Lewis [12]. It is hoped that the present discussion will stimulate further interest in this important problem of modern materials science.

REFERENCES

[1] Chandrasekhar, S, "Plasma Physics." Univ. of Chicago Press, Chicago, 1960.

[2] Pai, S., "Magnetogasdynamics and Plasma Dynamics." Springer, Berlin (Prentice-Hall, Englewood Cliffs, New Jersey), 1962.

[3] Spitzer, L., Jr., and Harm, R., Transport phenomena in a completely ionized gas, *Phys. Rev.* 89, 977 (1953).

[4] Burgers, J. M., Statistical plasma mechanics, *in* "Plasma Dynamics" (F. H. Clauser, ed.), p. 119. Addison-Wesley, Reading, Massachusetts, 1960.

[5] Born, M., and Green, H. S., "A General Kinetic Theory of Liquids." Cambridge Univ. Press, London and New York, 1949.

[6] Irving, J. H., and Kirkwood, J. G., The statistical mechanical theory of transport processes, IV. The equations of hydrodynamics, *J. Chem. Phys.* 18, 817 (1950).

[7] Hirschfelder, J. O., Curtiss, C. F., and Bird, R. B., "Molecular Theory of Gases and Liquids." Wiley, New York, 1954.

[8] Tolman, R. H., "Principles of Statistical Mechanics." Oxford Univ. Press, London and New York, 1938.

[9] Green, H. S., "Molecular Theory of Fluids." Interscience, New York, 1952.

[10] Kirkwood, J. G., The statistical mechanical theory of transport processes, I. *J. Chem. Phys.* 14, 180 (1946).

[11] Rostoker, N., and Rosenbluth, M. N., Fokker-Planck equation for a plasma with a constant magnetic field, *J. Nuclear Energy: Pt. C* 2, 195 (1961).

[12] Lewis, R. M., Solution of the equations of statistical mechanics. *J. Math. Phys.* 2, 222 (1961).

ON THE FLOW OF A CONDUCTING FLUID OF SMALL VISCOSITY†

Stephen Childress

California Institute of Technology
Pasadena, California

1. Introduction

As a result of the action of Maxwell stresses, the classical Bernoulli function $\mathcal{H}$ is in general not conserved in the stationary motion of an inviscid, electrically conducting fluid. Moreover, in an unbounded region $\mathcal{H}$ may be decreasing on streamlines and is not, in general, continuous at the point at infinity. This chapter is devoted to examples illustrating certain consequences of this particular property of inviscid flows, and to the possible significance of the difficulty at infinity in expansion procedures for flows of small viscosity.

It has long been realized that, owing to the possibility of Alfvén waves, a small perturbation of a given magnetohydrodynamic flow need not remain local or isotropic. It is in part this property of the fluid which accounts for the wake structure generated by a moving solid [1,2], and for the singular properties of the inviscid flows considered by Stewartson [3,4]. The variation of the Bernoulli function is likely to be significant, however, only when Joule heating occurs; that is, when the electrical conductivity of the fluid is finite and Alfvén waves are

†This paper presents the results of one phase of research carried out at the Jet Propulsion Laboratory, California Institute of Technology, under Contract No. NASw-6, sponsored by the National Aeronautics and Space Administration.

damped. Therefore, in the present investigation we shall be concerned not with the wave-like properties of magnetohydrodynamic flows, but with the effect of dissipation of currents and the combined effect of conductivity and viscosity upon this dissipation.

If the conductivity of the inviscid fluid is small, so that inviscid flows are influenced only slightly by some prescribed magnetic field, and do not seriously distort this field, Joule dissipation per unit volume is locally a small quantity. The variation of $\mathcal{H}$ may then be of no consequence in the first order inviscid theory, but will appear in a subsequent step in an expansion. The small variations in $\mathcal{H}$ behind a circular cylinder, arising in this way, have been described recently by Tamada [5]. Essentially the same remark applies also to the motion of a magnetized sphere [6], and in fact to a large class of axially symmetric flows [7].

If the interaction between the fluid and field is not small in the above sense, the inviscid theory is basically nonlinear. In the case of the uniform motion of a solid parallel to the field (Section 3), a nondiffusive wave is established, as in [5], but the disturbance of the uniform stream, or perturbation at infinity, is no longer small. An exact description of the perturbation cannot be given, but an expansion of the flow field, valid at large distances, provides an estimate of the asymptotic behavior of this wake near the outer edges, and an expression for its thickness in terms of the parameters of the problem. Also, the perturbation in velocity introduces into the formula for the force experienced by the solid, obtained by integral methods, a term expressing the finite flux of energy at infinity.

The diffusive effect which is provided by a small viscosity is examined in Section 4. There we consider the effect of a strong magnetic field on source and sink flows. We find that a continuous inviscid limit exists only in the case of sink flow. The source flow decays to zero at infinity only if the viscosity is finite. Therefore the source flow may be compared with the nondiffusive part of the flow field in the first example.

2. Inviscid Flows

We study the time-independent motion of an incompressible, electrically conducting fluid. According to the magnetohydrodynamic theory [8], if the viscosity of the fluid is identically zero, the relevant equations are

$$\mathbf{q} \cdot \nabla \mathbf{q} + \nabla p + \beta \mathbf{h} \times \mathbf{j} = 0 \tag{1a}$$

$$\mathbf{j} = \nabla \times \mathbf{h} = \mathrm{Rm}\,(\mathbf{e} + \mathbf{q} \times \mathbf{h}) \tag{1b}$$

$$\nabla \times \mathbf{e} = 0 \tag{1c}$$

$$\nabla \cdot \mathbf{q} = 0 \tag{1d}$$

$$\nabla \cdot \mathbf{h} = 0. \tag{1e}$$

All quantities in (1) are dimensionless and are defined by the transformations

$\mathbf{q}' = U\mathbf{q}$,	$\mathbf{q}'$	= velocity
$\mathbf{h}' = H\mathbf{h}$,	$\mathbf{h}'$	= magnetic field
$\mathbf{e}' = \mu U H \mathbf{e}$,	$\mathbf{e}'$	= electric field
$p' = \rho U^2 p + P$,	p'	= pressure
$\mathrm{Rm} = \sigma\mu\, UL$		= magnetic Reynolds number
$\beta = \dfrac{N}{\mathrm{Rm}}$,	$N = \dfrac{\mu^2 \sigma H^2 L}{\rho U}$	= interaction parameter

with U, H, P, and L representing reference values of velocity, magnetic field, pressure, and length, respectively.

We now write (1a) in the form

$$\nabla \mathcal{H} + \beta \mathbf{h} \times \mathbf{j} - \mathbf{q} \times \omega = 0; \qquad \mathcal{H} = p + \frac{1}{2} q^2, \qquad \omega = \nabla \times \mathbf{q}. \tag{2}$$

Integrating (1c) in terms of the electric potential ϕ, $\mathbf{e} = \nabla\phi$, we may deduce from (1), (2) the divergence relation

$$\nabla \cdot \mathbf{B} = -\frac{\beta}{\mathrm{Rm}} j^2 \leq 0; \qquad \mathbf{B} = \mathcal{H}\mathbf{q} - \frac{\beta}{\mathrm{Rm}} \phi \mathbf{j} \tag{3}$$

which exhibits on the right-hand side minus the rate at which energy is dissipated by Joule heating, per unit volume. It is an immediate consequence of (3) that steady inviscid flows defined in a *bounded* region, whose boundary is insulated and across which there is no flow are not dissipative. For, according to the hypotheses the normal components of $\mathbf{q}$ and $\mathbf{j}$ vanish on the boundary, and therefore by (3) $j = 0$ everywhere. However, if an unbounded region is considered (as in the motion of a solid through an infinite volume of fluid), the flow may be dissipative and in particular $\mathcal{H}$ can vary along streamlines. This variation is such, however, that $\mathcal{H}$ is constant of trajectories of the vector field $\mathbf{A}$, where

$$\mathbf{A} = \frac{\mathbf{q}[\boldsymbol{\omega} \cdot (\mathbf{q} \times \mathbf{h})] - \beta \mathbf{h}[\mathbf{j} \cdot (\mathbf{q} \times \mathbf{h})]}{|\mathbf{q} \times \mathbf{h}|} .$$

This remark follows from (2).

We restrict attention now to steady two-dimensional and axially symmetric flows. In these flows the streamlines and magnetic field lines lie in the (x,y) plane, and the electric field vanishes. The following notation is used:

$$\mathbf{q} = u\mathbf{i} + v\mathbf{j}, \qquad \mathbf{h} = h_1\mathbf{i} + h_2\mathbf{j}$$

$$\omega = \frac{\partial v}{\partial x} - \frac{\partial u}{\partial y} = \text{vorticity}, \qquad j = \frac{\partial h_2}{\partial x} - \frac{\partial h_1}{\partial y} = \text{current}.$$

We also make use of the integral parameter m, $m = 0$ for two-dimensional flow, $m = 1$ for axially symmetric flow. The divergence relation (3) becomes

$$\mathbf{q} \cdot \nabla \mathcal{H} = -\frac{\beta}{\text{Rm}} j^2 ; \tag{4}$$

that is, for these symmetrical flows the Bernoulli function decreases on streamlines. Also, the Bernoulli function becomes the *streamfunction* for the vector $y^{-m}\mathbf{A}$,

$$\frac{1}{y^m} \frac{\partial \mathcal{H}}{\partial y} = \frac{1}{y^m} (\beta h_1 j - u\omega)$$

$$-\frac{1}{y^m} \frac{\partial \mathcal{H}}{\partial x} = \frac{1}{y^m} (\beta h_2 j - v\omega).$$

By (1d) the velocity also has a streamfunction ψ, and it follows that for flows defined in the simply connected region R of the

(x,y) plane, bounded by the contour C, the effect of dissipation is expressed by

$$\frac{\beta}{\mathrm{Rm}} \iint_R y^m j^2\, dx\, dy = \iint_R \left(\frac{\partial \psi}{\partial x} \frac{\partial \mathcal{H}}{\partial y} - \frac{\partial \mathcal{H}}{\partial x} \frac{\partial \psi}{\partial y} \right) dx\, dy$$

$$= \oint_C \psi\, d\mathcal{H} = - \oint_C \mathcal{H}\, d\psi \geq 0. \tag{5}$$

3. The Uniform Motion of a Finite Solid

If a two-dimensional or axially symmetric solid moves with uniform speed parallel to a magnetic field equal to $H\mathbf{i}$ at infinity, the inviscid flow satisfies the equations given above and the boundary conditions (we omit the surface conditions on the electromagnetic field)

$$h_1 = 1, \qquad h_2 = v = p = 0 \qquad \text{at infinity} \tag{6a}$$

$$\mathbf{q} \cdot \mathbf{n} = 0 \qquad \text{at the solid.} \tag{6b}$$

The condition

$$\mathbf{q} = \mathbf{i} \qquad \text{at infinity} \tag{7}$$

is not now applicable, unless $j = 0$ everywhere in the region. This is because, if (6) is satisfied, and if the velocity is continuous at infinity and satisfies there (7), then $\mathcal{H}$ is continuous at infinity and, by (5), $j = 0$ on every streamline.

Ruling out the exceptional, nondissipative solutions (which are in general not relevant), the condition at infinity which is applied here, in place of (7), requires that $\mathbf{q}$ converge to a limit arbitrarily close to $\mathbf{i}$, as (x,y) tends to infinity along any path that does not lie in the strip S consisting of the points $x \geq 0$, $|y| < M$, provided that M is sufficiently large. It will be assumed also that there exists a nonnegative function $F(y;\beta;\mathrm{Rm})$ such that

$$\mathbf{q} \to \mathbf{i} - \mathbf{i}F(y;\beta;\mathrm{Rm}) \qquad \text{as} \quad x \to \infty, \quad y \text{ fixed.} \tag{6c}$$

The function F expresses the possibility of a decrease in $\mathcal{H}$ along each streamline. The disturbance near the positive

x-axis represents a nondiffusive wake whose structure is at this point unknown. There is also a certain arbitrariness in condition (6c); perturbations in p and $\mathbf{q}$ are equally possible in principle, as are perturbations on the negative x-axis. We shall return to this point later.

With the boundary condition at infinity modified as described above, Tamada [5] constructed the function F for two-dimensional motion past a circular cylinder, assuming the conductivity of the fluid to be small. We shall relax the last condition and instead restrict attention to the flow field at large distances and to the behavior of $F(y)$ for large values of the argument. Our basic hypothesis is that an expansion of the solution outside S can be accomplished by iteration, starting with the free stream values. This leads to expansion in terms of solutions of linear equations. For two-dimensional flow past a circular cylinder, we have the expansion [9]

$$\mathbf{q} = \mathbf{i}\left(1 + Nk^{(1)}\right) - \frac{1}{\beta - 1}\nabla k^{(2)} + o(k^{(1)}) + o(r^{-1}) \tag{7a}$$

$$\mathbf{h} = \mathbf{i}\left(1 + \mathrm{Rm}k^{(1)}\right) - \frac{1}{\beta - 1}\nabla k^{(2)} + o(k^{(1)}) + o(r^{-1}) \tag{7b}$$

$$p = -u + o(k^{(1)}) + o(r^{-1}) \tag{7c}$$

where†

$$k^{(1)} = -\frac{\gamma}{2\pi} e^{-\frac{1}{2}(N-\mathrm{Rm})x} K_0\left(\frac{|N - \mathrm{Rm}|}{2} r\right),$$

$$k^{(2)} = \frac{\gamma}{2\pi} \log r, \qquad r = \sqrt{x^2 + y^2} \tag{7d}$$

and γ is a constant. It follows from (1b) that

$$j = -\mathrm{Rm}\frac{\partial k^{(1)}}{\partial y} + o\left(\frac{\partial k^{(1)}}{\partial y}\right) \tag{7e}$$

and the leading term in the expansion of $\mathcal{H}$ outside the strip S may be found from (4) with the boundary condition at infinity,

†It is assumed here that $\beta \neq 1$.

using (7e) (see Fig. 1). The expansion of F then follows and, enlarging our results to include also axially symmetric flow past a sphere, it is found that

$$F(y;\beta;\mathrm{Rm}) = \frac{\gamma^2}{2(2\pi)^{m+1}} \frac{\beta}{|1-\beta|} y^{-(2m+2)} + o(y^{-(2m+2)}) . \tag{8}$$

A measure of the width of this wake is therefore given by

$$\delta = \left(\frac{\beta\gamma^2}{|1-\beta|}\right)^{1/(2m+2)} \tag{9}$$

The constant γ is equal to the total instantaneous mass flux in the wake relative to the free stream, and is defined by

$$\gamma = \int (2\pi y)^m F(y)\, dy , \tag{10}$$

integration being over the plane normal to the direction of motion. It may also be shown that if $0 \leq F \leq 1$ uniformly in y, then $0 < \gamma \leq D$, where D is the dimensionless drag experienced by the obstacle. Thus δ may be bounded from above in terms of D and β.

The proof of this last assertion rests upon the validity of (7) on portions of a distant contour where the Maxwell stresses are small. In two dimensions we have

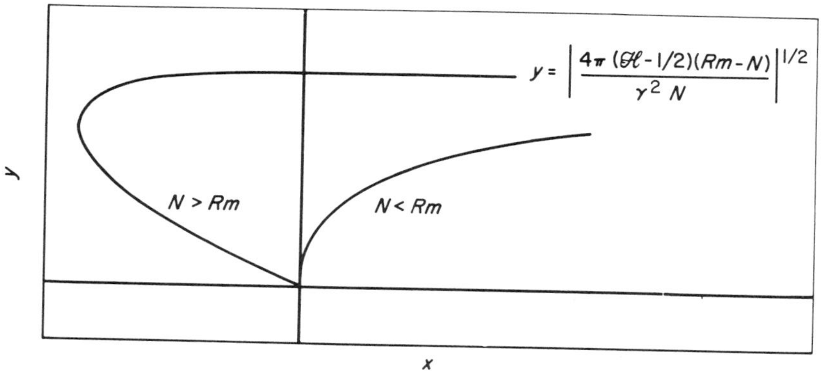

Fig. 1. Lines of constant ℋ at large distances, two-dimensional flow.

$$D = \int_C \left[\mathbf{q}(u-1) - \beta\mathbf{h}(h_1 - 1) + \mathbf{i}\left(p + \frac{1}{2}\beta h^2\right)\right] \cdot \mathbf{n}\, ds \tag{11}$$

where C is a contour containing the cylinder, and $\mathbf{n}$ is its outward normal. In order to evaluate the contour integral, C is taken to consist in part of the segment $x = x_0 > 0$, $|y| < M$; we may then use (7) to show that D is equal to the integral over the wake. The same statement applies to the axially symmetric flow. We find therefore that

$$D = \int (2\pi y)^m \,(F - F^2)\, dy \tag{12}$$

which establishes the result.

We remark that (12) may also be proved by noting that the rate at which energy is converted into heat may be calculated from (5), (7), and for the infinite fluid is equal to

$$\int (2\pi y)^m \,(1 - F)\left(F - \frac{1}{2}F^2\right) dy\,. \tag{13}$$

In addition, kinetic energy is transferred to the fluid through the point at infinity at the rate

$$\int (2\pi y)^m \,\frac{1}{2} F^2 (1 - F)\, dy\,. \tag{14}$$

The sum of (13) and (14) must equal the work done on the fluid by the obstacle, per unit time, and (12) is obtained.

Although the results described above are derived under the strict assumption that the fluid is inviscid, it is of interest to know the relation over a fixed finite region between these solutions and a limit obtained from a one-parameter family of solutions, depending on viscosity, as the latter tends to zero. It is known that if viscosity is small but nonzero, the asymptotic structure of wakes caused by the motion of a solid parallel to the field is similar to the ones given here [1,2], but in that theory vorticity diffuses (slowly), essentially at constant pressure, as it is convected toward $x = +\infty$. It would appear therefore that inviscid solutions which are limits in the above sense are of the type discussed above; that is, the pressure is continuous and the velocity satisfies (6c).

A reasonable justification for this conclusion, however, must consider more precisely the decay of the disturbance at $x = +\infty$ under the action of viscosity. In particular the speed of convection is a function of y, so the diffusion is not of a simple type. We propose instead to examine the interplay between diffusion of vorticity by viscosity, and diffusion of current by electrical conductivity, by means of a second example which, however, shares certain properties with the dual wake structure which may be associated with a moving solid.

4. The Effect of Small Viscosity on Source and Sink Flows

We adopt now, in place of (1a), the full Navier-Stokes equation

$$\mathbf{q} \cdot \nabla \mathbf{q} + \nabla p + \beta \mathbf{h} \times \mathbf{j} - \frac{1}{Re} \nabla^2 \mathbf{q} = 0. \qquad (1a')$$

The discussion is again restricted to two-dimensional and axially symmetric flow, and the notation of Section 3 is used. The following problem is considered. Let the fluid lie in the region $x > 0$, and suppose that the magnetic field is equal to $H\mathbf{i}$ at infinity. The conductivity of the material filling the region $x < 0$ is assumed finite. The flow is due to the removal of mass at the origin at the rate Q, and may be said to constitute a sink flow. We also consider source flows, where Q is the rate at which mass is added to the fluid. For these flows the physical parameters L, U, and Q are not independent and are related by

$$\rho U L^{m+1} = Q . \qquad (15)$$

The Navier-Stokes equations are to be replaced by a simpler, but still nonlinear, system which is obtained from (1a') by a limiting procedure. It is assumed that the parameter H is large, so that

$$N \gg 1, \qquad \beta \gg 1. \qquad (16)$$

Because N is large, the field $H\mathbf{i}$ largely controls the motion of the fluid, and the curvature of streamlines is small; this leads to a boundary layer approximation. If in addition β is large, the magnetic field may be replaced by the value at infinity, in the construction of the velocity and pressure of the flow; the magnetic field perturbation appears at a later stage [9]. The approximate system which emerges from the formal limit process involves a set of new variables, defined here by an overbar, which are related to the old variables by the equations

$$\bar{u} = u, \qquad \bar{x} = x/N$$

$$\bar{v} = Nv, \qquad \bar{y} = y.$$

$$\bar{p} = p$$

The equations for $\bar{u}$, $\bar{v}$, and $\bar{p}$ are

$$\bar{u}\frac{\partial \bar{u}}{\partial \bar{x}} + \bar{v}\frac{\partial \bar{u}}{\partial \bar{y}} + \frac{\partial \bar{p}}{\partial \bar{x}} - \kappa\left(\frac{\partial^2 \bar{u}}{\partial \bar{y}^2} + \frac{m}{\bar{y}}\frac{\partial \bar{u}}{\partial \bar{y}}\right) = 0 \tag{17a}$$

$$\frac{\partial \bar{p}}{\partial \bar{y}} + \bar{v} = 0 \tag{17b}$$

$$\frac{\partial \bar{u}}{\partial \bar{x}} + \frac{\partial \bar{v}}{\partial \bar{y}} + \frac{m\bar{v}}{\bar{y}} = 0 \tag{17c}$$

and contain the viscosity in the dimensionless parameter

$$\kappa = \frac{\nu\mu^2\sigma H^2}{\rho U^2}, \qquad \nu = \text{kinematic viscosity.}$$

The system (17) is to be regarded now as defining, together with certain conditions, an "exact" solution, or rather a family of exact solutions depending on viscosity, and we ask for approximations valid in the asymptotic sense for small κ. In the exact sink-source problem, mass is transferred at a point and the parameters U and L must therefore be eliminable in favor of Q. This dimensional constraint implies that the exact solutions have the following similarity form:

$$\bar{u} = \kappa^{1/2}\, u^*(x^*, y^*) \tag{18a}$$

$$\bar{v} = \kappa^{(2m+3)/[2(m+1)]}\, v^*(x^*, y^*) \tag{18b}$$

$$\bar{p} = \kappa p^*(x^*, y^*) \tag{18c}$$

where

$$x^* = \kappa^{(3-m)/2}\,\bar{x} \tag{18d}$$

$$y^* = \kappa^{1/(2m+2)}\,\bar{y}\,. \tag{18e}$$

The boundary conditions imposed upon u^*, v^*, and p^* are

$$\int_{x^*>0} (2\pi y)^m\, u^*\, dy^* \pm 1 = 0 \tag{19a}$$

$$u^* = v^* = p^* = 0 \quad \text{at infinity} \tag{19b}$$

$$u^* \to 0 \quad \text{as} \quad x^* \to 0, \quad y^* > 0, \tag{19c}$$

the upper sign in (19a) corresponding to sink flow.

Purely on the basis of the similarity (18) of exact solutions, it may be shown that the viscosity of the fluid, however small, will ultimately have an effect upon the *decay* of source and sink flows. For, suppose that an approximation $\bar{u}_1$ to $\bar{u}$ is found which is valid to order $\kappa^{1/2}$ inclusive uniformly in $\bar{x} \geq \epsilon \geq 0$, where ϵ is arbitrary; that is, we assume that for any $\epsilon_1 > 0$, there is a κ_0 such that

$$\frac{|\bar{u} - \bar{u}_1|}{\kappa^{1/2}} < \epsilon_1, \qquad \kappa < \kappa_0$$

uniformly. Then, the limit of $\kappa^{-1/2}\bar{u}_1$ as κ tends to zero, x^*, y^* fixed, and $x^* > 0$, exists and equals $u^*(x^*, y^*)$.† This is proved by noting that, by the similarity (17),

$$|u^*(x^*, y^*) - u_1^*(x^*, y^*; \kappa)| < \epsilon_1, \qquad \kappa < \kappa_0$$

uniformly for $x^* \geq \kappa^{(3-m)/2}\epsilon$, where we have defined $\kappa^{-1/2}\bar{u}_1 = u_1^*(x^*, y^*; \kappa)$. But for any fixed x^* there is a $\kappa_1 < \kappa_0$ such that x^* exceeds $\kappa_1^{(3-m)/2}\epsilon$.

†Cf. [10], p. 821.

The effect of viscosity depends, however, on whether the flow is a sink flow or a source flow. In fact, the inviscid limit u', v', p' defined by

$$u' = \lim_{\kappa \to 0} \bar{u}, \qquad \bar{x}, \bar{y} \text{ fixed, etc.,}$$

is continuous in $x \geq \epsilon$ only in the case of sink flow. The continuous sink flow was found for the axially symmetric case by Kovásznay and Fung [11] and independently for the two-dimensional and axially symmetric cases by the writer. The equations are

$$u' \frac{\partial u'}{\partial \bar{x}} + v' \frac{\partial u'}{\partial \bar{y}} + \frac{\partial p'}{\partial \bar{x}} = 0 \tag{20a}$$

$$\frac{\partial p'}{\partial \bar{y}} + v' = 0 \tag{20b}$$

$$\frac{\partial u'}{\partial \bar{x}} + \frac{\partial v'}{\partial \bar{y}} = 0. \tag{20c}$$

Since L must be eliminable, the inviscid flow has in either case the form

$$u' = \frac{1}{\bar{y}^m} \frac{\partial \psi'}{\partial \bar{y}}, \qquad v' = -\frac{1}{\bar{y}^m} \frac{\partial \psi'}{\partial \bar{x}}, \qquad \psi' = \operatorname{sgn} \bar{y}\, f(\xi)$$

$$p' = \bar{x}^{2/(m-3)}\, g(\xi), \qquad \xi = \frac{\bar{y}^2}{a_m \bar{x}^{2/(3+m)}} \tag{20d}$$

where $a_0 = 3\sqrt{2}$, $a_1 = 8$. The equation for g follows upon substitution in (20) and is

$$\left(\frac{dg}{d\xi}\right)^2 + \xi \frac{dg}{d\xi} + (m+1)g = 0. \tag{21}$$

We may integrate (21) by noting that

$$\frac{1}{4} - (m+1)\frac{g}{\xi^2} \geq 0.$$

With the substitution

$$\phi = \left[\frac{1}{4} - (m+1)\frac{g}{\xi^2}\right]^{1/2} + \frac{1}{2} \tag{22}$$

there is obtained a polynomial for ϕ of degree $m + 3$

$$\left[(2-m)\phi + 1\right]^{m+2} (\phi - 1) = \frac{b_m}{\xi^{m+3}} . \tag{23}$$

The constant b_m is fixed by the condition on g following from (19a) and for positive b_m a sink flow results [9]. If $b_m \le 0$ the only continuous solution is $g = 0$. This follows from (23) and refers directly to the effect of dissipation. The quantity

$$\bar{\mathfrak{H}} = \bar{p} + \frac{1}{2}\bar{u}^2$$

replaces the function $\mathfrak{H}$ in these solutions, and it is seen from (20) that for inviscid flows $\bar{\mathfrak{H}}$ is decreasing on streamlines (see Fig. 2). Now

$$\mathfrak{H}' = -\frac{1}{(m+1)\,\bar{x}^{2/(3-m)}}\,\xi\,\frac{dg}{d\xi}$$

and is opposite in sign to $dg/d\xi$. Suppose that $b_m \le 0$; then $\phi - 1 \le 0$ from (23), except possibly when $m = 1$ and $\phi + 1 \le 0$, but this means still that $\phi - 1 \le 0$. Thus, from (21), (22), and (23)

$$\frac{dg}{d\xi} = 2\xi(\phi - 1) \le 0. \tag{24}$$

On the other hand, $g = O(\xi^{-m-1})$ as $\xi \to \infty$ and

$$\int_0^\infty \xi^{(m-1)/(2-m)}\left[\left(\frac{dg}{d\xi}\right)^2 + \frac{1}{2-m}\,g\right] d\xi = -\int_0^\infty \frac{d}{d\xi}\left(\xi^{1/(2-m)}\,g\right) d\xi = 0$$

$$= \frac{1}{2}\int_0^\infty \xi^{(m-1)/(2-m)}\left[\left(\frac{dg}{d\xi}\right)^2 - \xi\,\frac{dg}{d\xi}\right] d\xi$$

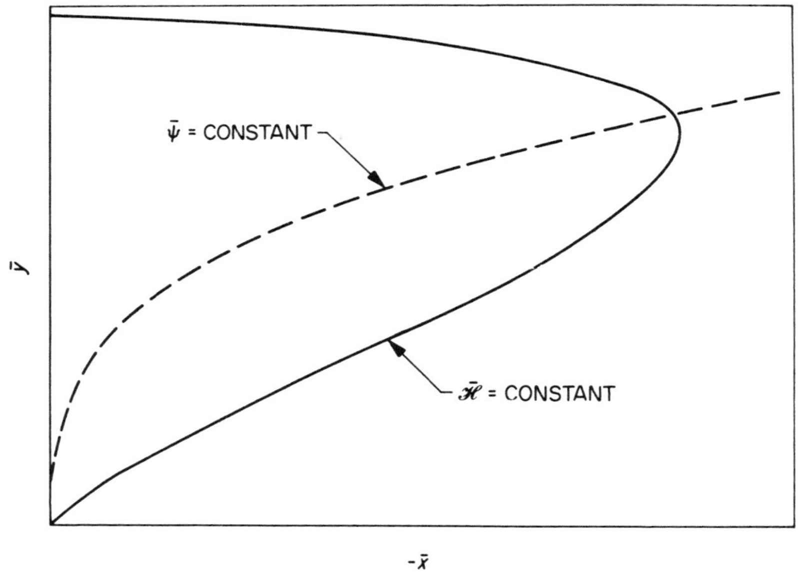

Fig. 2. Variation of $\bar{\mathcal{H}}$ for inviscid, two-dimensional sink flow.

which because of (24) is possible only if $dg/d\xi = 0$, $\xi \geq 0$; if $g = 0$ at $\xi = \infty$, this means that $g = 0$, $\xi \geq 0$.

The effect of *small* viscosity is, therefore, not strong in the sink flow, and may be neglected except at very large distances, where the perturbation in velocity is already small. However, in the source flow it is only through the diffusion of vorticity that the required similarity is attained. We can construct *inviscid* source flows, continuous over a finite region, which satisfy (19a), but they will contain a length.

REFERENCES

[1] Hasimoto, H., *Revs. Modern Phys.* 32, 860-867 (1960).
[2] Gourdine, M. C., *J. Fluid Mech.* 10, 439-448 (1961).
[3] Stewartson, K., *J. Fluid Mech.* 8, 82-96 (1960).
[4] Stewartson, K., *Revs. Modern Phys.* 32, 855-859 (1960).

[5] Tamada, K., *Phys. Fluids* 5, 817-823 (1962).
[6] Ludford, G. S. S., and Murray, J. D., *J. Fluid Mech.* 7, 516-528 (1960).
[7] Childress, S., Jet Propulsion Lab., Space Programs Summary No. 37-15, p. 125 (1962).
[8] Cowling, T. G., "Magnetohydrodynamics." Interscience, New York, 1957.
[9] Childress, S., Jet Propulsion Lab., TR 32-351 (January 1963).
[10] Chang, I-Dee, *J. Math. and Mech.* 10, 811-876 (1961).
[11] Kovásznay, L. S. G., and Fung, F. C. W., *Phys. Fluids* 5, 661-664 (1962).

ANALOGY OF MAGNETOHYDRODYNAMIC AND PLASTIC FLOWS

Tsung Lien Chou
Michigan College of Mining and Technology
Houghton, Michigan

Introduction

In the ever widening field of the magnetohydrodynamics, as in ordinary fluid mechanics, two main divisions may be obtained, namely the compressible and the incompressible. At the present moment, the outlook of the former seems broader as it includes such important problems of atomic power from fusion reactors, hypersonic flights, guiding and controlling rockets and missiles, and the stellar evolutions especially the phenomenon of supernova. On the other hand, the application of the incompressible is involved only in such limited areas of metal refining, pumping of conducting liquids and central-station power generation. However, the great challenge comes from the power development for astronautic navigation into deep space for duration of years. For such conditions the possible solution for propelling power may come from the atomic pile through conducting liquid accelerated by gas through a nozzle. In the preliminary stage of research, liquid lithium has been tried with a promise far above the thermionic or the thermio-electronic processes. For successful design and operation of plants involving the flow of conducting liquids, data are urgently needed for velocity profile, friction factor, tractive resistance, and the criteria for controlling types of flow.

Theoretically, these problems are interwoven by electromagnetic phenomena with that of the hydrodynamics, mutually modified. In other words, what amounts to is the simultaneous solution of sixteen equations of Maxwell and Navier-Stokes. At this stage of development of the science, nothing practical can be expected, especially for cases involving nonlinearity. Alternatively, the phenomenological process may shed some useful light. Along this line, some efforts have been made to obtain some information for engineering design. Among the liquids used are mercury, sodium potassium alloy, and liquid lithium. The phenomena are complicated by the combined influences and furthermore by the effects of the relative strengths of the applied magnetic fields and the one induced by flowing fluid. Thus basic problems remain to be cleared up.

On the other hand, there are some liquid flows having similar effects in chemical and processing industries. These include various sludges, slurries, mixtures, and solutions which modify the hydrodynamic flow by additional friction loss and changing the velocity distribution due to causes other than the hydrodynamics. However, one point should be mentioned here. These flows are called non-Newtonian, including some plastic and dilatant effects in various degrees. But among the whole group of the non-Newtonian flows, only Bingham plastic is taken as the representative for the following reasons: the dilatant and the pseudoplastic flows are characteristic in limited ranges and they may approach either the Newtonian or the plastic in approximate treatment. Secondly, more experimental data are available in this area.

Basic Relations

The significant characteristics of the Bingham plastic flow is that the flow starts only when the unit shearing stress is above certain initial value called the yielding stress τ_y and from that point onward, the shearing stress is proportional to the shearing rate du/dy with a constant ratio η, the coefficient of rigidity in the complete laminar range. This is illustrated on the paragenic diagram of τ versus du/dy together with the Newtonian and other non-Newtonian fluids. The common expression is

$$\tau - \tau_y = \eta(du/dy). \tag{1}$$

This is shown in Fig. 1. Apparently, the effect of the exotic material on the pure liquid is to increase the resistance to the flow by the initial obstacle and also the higher value of coefficient of viscosity. So far, it has been found that this modifying influence varies with the shearing rate.

When a conducting fluid moves at velocity $\mathbf{u}$ across a magnetic field $\mathbf{B}$, an electric current $\mathbf{j}$ is induced as

$$\mathbf{j} = \sigma \mathbf{u} \times \mathbf{B}$$

in a direction normal to both vectors, where σ is the electric conductivity. In turn, this current $\mathbf{j}$ reacted by the magnetic field $\mathbf{B}$ produces an electromagnetic force $\mathbf{F}$ (see Fig. 2)

$$\mathbf{F} = \mathbf{j} \times \mathbf{B} = \sigma \mathbf{u} \times \mathbf{B} \times \mathbf{B}$$

in a direction normal to vector $\mathbf{j}$ and $\mathbf{B}$, that is, just opposite to $\mathbf{u}$ (called Lorentz or preponderomotive force). Then

$$\mathbf{F} = \sigma(\mathbf{u} \times \mathbf{B}) \times \mathbf{B} = \sigma\left[\mathbf{B}(\mathbf{u} \cdot \mathbf{B}) - \mathbf{u}(\mathbf{B} \cdot \mathbf{B})\right] .$$

Since $\mathbf{B}$ is normal to $\mathbf{u}$, hence $\mathbf{u} \cdot \mathbf{B} = 0$. Therefore

$$\mathbf{F} = -\sigma B^2 \mathbf{u}. \tag{2}$$

As soon as the fluid flows into the magnetic field, this force acts on every unit mass to decelerate the flow from an initial velocity $\mathbf{u}$ to $\mathbf{u}_m$ in a distance L, or

$$\mathbf{F} = (\mathbf{u}_m - \mathbf{u})/L = \sigma \mathbf{u} B_0^2 .$$

Therefore

$$\begin{aligned} \mathbf{u}_m &= \mathbf{u}(1 - \sigma L B^2/\rho) \\ &= \mathbf{u}(1 - \lambda) \end{aligned} \tag{3}$$

where $\lambda = \sigma B_0 L/\rho$ is a factor representing the interaction of the magnetic field on fluid flow.

If both sides of (3) are differentiated with respect to y, the coordinate normal to the solid boundary, we have

$$d\mathbf{u}_m/dy = (1 - \lambda)(d\mathbf{u}/dy).$$

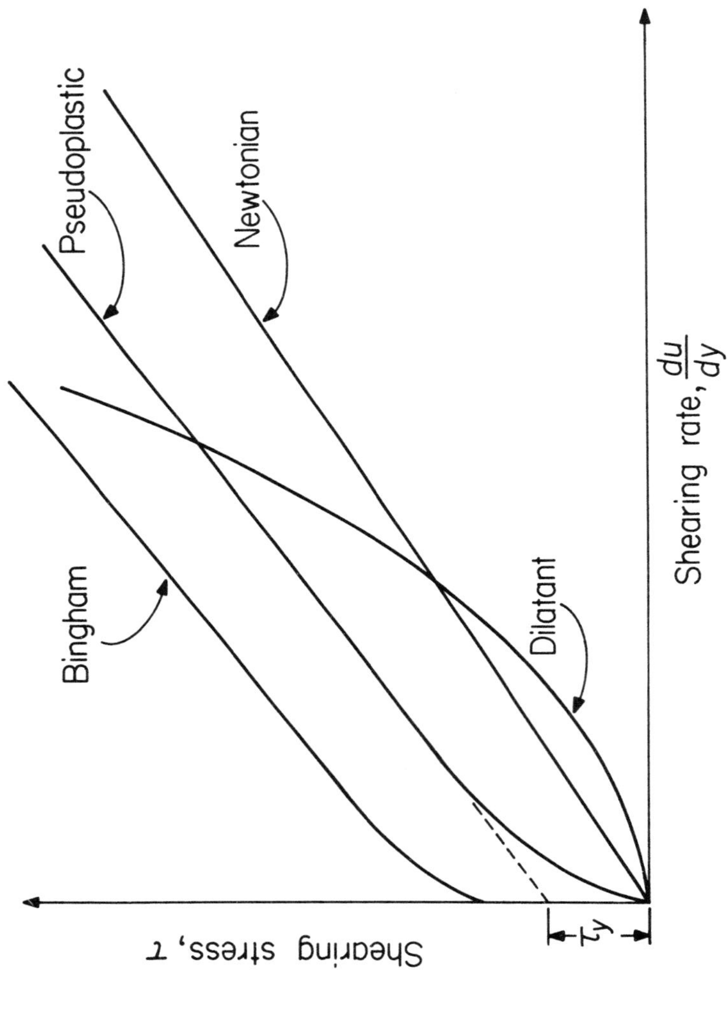

Fig. 1. Hydrodynamic flow.

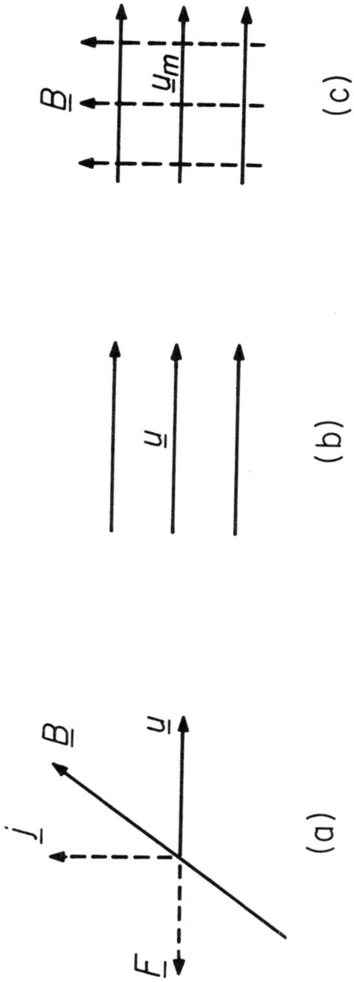

Fig. 2. Magnetohydrodynamic flow: (a) vectors; (b) hydrodynamic; (c) magnetohydrodynamic.

If Newtonian relation of viscosity is applied, the unit shear in the fluid is

$$\tau_m = \mu_m(du/dy)$$

$$= \mu_m(du_m/dy)\,[1/(1-\lambda)]$$

or

$$\tau_m - \lambda\tau_m = \mu_m(du_m/dy). \quad (4)$$

It is evident that the structures of (1) and (4) are quite similar in the basic idea of shearing stress. The interesting deductions are that, for each magnetic field strength **B** and flow, there is a magnetic-flow viscosity μ_m which may be constant in certain range of laminar flow; and there is an initial shearing stress to start the flow. However, it should be noted that λ varies with the flowing velocity **u** also.

Equation (4) also indicates the functional relation of **B**, σ, and **u** and experimental information can be obtained by studies in that direction. Besides plot of such results in the form shown in Fig. 1, it may shed some basic light on the stability of laminar flow. The other points to be noticed is that the so-called Bingham plastic is an ideal fluid to be approximated by actual fluid and so will be the hydromagnetic flows.

Skin Friction over Flat Plate

In a pure hydrodynamic flow, the coefficient of skin friction C_f is defined as the unit shearing stress τ to the dynamic pressure $\rho u^2/2$ or

$$C_f = \tau/(\rho u^2/2). \quad (5)$$

In flows other than pure hydrodynamic, the apparent or corresponding shearing stress $\mu(d\mathbf{u}/dy)$ should be used in evaluating the coefficient in Newtonian way.

In plastic flow, the coefficient is

$$C'_f = \frac{\eta(du/dy)}{\rho u^2/2}.$$

From Eq. (1) and the hydrodynamic shearing stress

$$\tau = \mu(d\mathbf{u}/dy)$$

we have

$$C'_f = \frac{\mu(du/dy) - \tau_y}{\rho u^2/2} . \tag{6}$$

From the two preceding equations, the ratio of the coefficients becomes

$$\frac{C'_f}{C_f} = 1 - \frac{\tau_y}{du/dy} . \tag{7}$$

The ratio will be reduced as the yielding stress τ_y increases. This does not mean the total drag or the unit shearing stress is smaller for plastic flow of higher yielding stress. The simple explanation is that the final shearing stress is evaluated in terms of μ of Newtonian flow which is almost always greater than η. This is the point where some confusion and controversy have arisen.

In magnetic flow, the coefficient is

$$C_{fm} = \frac{\mu_m(du_m/dy)}{\rho u^2/2} .$$

If μ_m is approximated by μ, the ratio of the coefficients is

$$C_{fm}/C_f = 1 - \lambda. \tag{8}$$

Friction Factor

The Darcy-Weisbach's friction factor f is defined as

$$\tau_0 = \frac{f}{4}\frac{\rho u^2}{2} . \tag{9}$$

where τ_0 is the shearing stress at wall surface. In the laminar flow of Newtonian fluid, it is given as

$$f = 64/R \tag{10}$$

where $R = \rho uD/\mu = uD/\nu$ is the Reynolds number.

In plastic flow, the pressure drop P may depend on density ρ, rigidity η, yielding stress τ_y, mean velocity u, and some characteristic length D. By the principle of dimensional analysis, the final functional relation is

$$f = \phi\left(\frac{uD\rho}{\eta}, \frac{\tau_y D^2}{\eta^2}\right). \tag{11}$$

Here $R = \rho uD/\eta$ for the plastic flow, and

$$H_E = \frac{\tau_y \rho D^2}{\eta^2} \tag{12}$$

is the Hestrom number which stands for the effect of yielding stress.

From pipe flow analysis [1], we obtain

$$\frac{8u}{D} = \frac{\tau_0}{\mu_p}\left[1 - \frac{4}{3}\frac{\tau_y}{\tau_0} + \frac{1}{3}\left(\frac{\tau_y}{\tau_0}\right)^4\right].$$

By making $\mu_p = \eta$ and taking τ_0 from Eq. (9), we arrive at

$$\frac{8u}{D} = \frac{\rho u^2 f}{8\eta}\left[1 - \frac{4}{3}\frac{\tau_y}{\tau_0} + \frac{1}{3}\left(\frac{\tau_y}{\tau_0}\right)^4\right].$$

But $R = \rho uD/\mu = \rho uD/\eta$ approximately, hence

$$\frac{1}{R} = \frac{f}{64}\left[1 - \frac{4}{3}\frac{\tau_y}{\tau_0} + \frac{1}{3}\left(\frac{\tau_y}{\tau_0}\right)^4\right]. \tag{13}$$

However,

$$\tau_y/\tau_0 = (8/f)(H_E/R^2), \tag{14}$$

Therefore

$$\frac{1}{R} = \frac{f}{64} - \frac{H_E}{6R^2} + \frac{1}{3f^3}\,\frac{{H_E}^4}{R^8} \;. \tag{15}$$

If the last term is neglected because of the high power of a quantity less than unit, the final equation is

$$f = \frac{64}{R} + \frac{32}{3}\left(\frac{H_E}{R^2}\right) \tag{16}$$

or

$$f = \frac{64}{R} + \frac{4}{3}\left(\frac{\tau_y}{\tau_0}\right) \;. \tag{16a}$$

Thus the plastic effect is additive and the relation becomes hydrodynamic, if the yielding stress approaches zero, provided the plastic Reynolds number is used.

On hydromagnetic flow, the laminar flow friction is derived by Hartmann as

$$f = \frac{32}{R}\,\frac{M^2 \tanh M}{M - \tanh M} \tag{17}$$

for mercury flowing through a normal magnetic field **B** where $R = \rho D_H \mathbf{u}/\eta$, D_H = twice the channel width, ρ and η are the unit mass and viscosity of mercury. $M = BL\sqrt{\sigma/\eta}$ is called the Hartmann number. This formula is proposed to fit into experimental data and can be reduced into

$$f = \frac{R}{64}\left[\frac{\tanh^2 M}{2} + \frac{M \tanh M}{2} + \frac{\frac{1}{2}\tanh^2 M}{M - \tanh M}\right] . \tag{18}$$

When the magnetic field **B** approaches zero, the friction factor should be changed over into that of hydrodynamic flow $f = R/64$. But the value in the bracket of (18) is not unit when M is zero. It may be inferred that the transcendentical function is not well checked with the physical fact, though fitted in the data.

The remarkable correspondence of the two kinds of flows is the more or less fixed values of f at transition from laminar to turbulent. For plastic flow, Metzner estimates the critical R at 2100 to 2800 and f at 0.032. This falls within the range of other experiments.

In hydromagnetic flow, Harris maintains, from the experimental data of Hartmann and Murgatroyd, that the critical f_c falls in the range of $0.035 < f_c < 0.044$. The values agree closely with each other.

Velocity Profiles

In plastic flow, the yielding value plays prominent role in shaping the velocity profile. In laminar stage, the velocity distribution from the wall to the point r_y of τ_y is

$$u = \frac{1}{\eta}\left(\frac{r_0\tau_0}{2} - \frac{r^2\tau_0}{2r_0} - \tau_y r_0 + \tau_y r\right) \tag{19}$$

up to u_r. For the central portion, the whole thing moves along like a solid plug with the constant velocity

$$u_y = \frac{1}{\eta}\left(\frac{r_0\tau_0}{2} - \frac{r_y^2\tau_0}{2r_0} - \tau_y r_0 + \tau_y r_y\right) . \tag{20}$$

This is the general velocity profile of ideal plastic flow in the laminar stage. In the practical plastic flow, the picture may be slightly different from this.

Now, when the turbulence sets in, what is the modification of the effect of yielding value? So far, testing results are contradictory and confusing. The main reason is that the transition is varying itself with the velocity and turbulence intensity. In the initial step of turbulent flow, the regions between the solid wall and the central plug becomes turbulent with transition on each side. The central core may start on laminar flow, not moving as solid plug any more. As the degree of turbulence advances, the two saddle turbulent regions will push toward the central region, but there the turbulent shearing stresses are still much smaller than in the two saddle cores. Such velocity distribution is borne out by experiments of Ismail, and Vanoni and Nomicos on sand, Babbitt and Caldwell on sludges and slurries. Even at very turbulent flows, the velocity in central portion is more or less uniform and the fluctuations are also less. In other words, the addition of some solid particles into the fluid has the general damping effect on eddies, especially near the solid boundaries and the central core.

Quite similarly, the magnetic field has a damping effect on the turbulence in a channel flow, when they are normal to each other. In the range of $R = 500$ to $R = 10{,}000$, the following outline may be made. With a small magnetic field, the general damping effect on the random eddies causes the increase in velocity itself and velocity gradient all over the section and the total discharge is increased when the pressure drop is kept constant. When the field is increased, the central core is reduced to a laminar flow and the discharge will be decreased. Further increase in the magnetic field changes the central portion into a plug with uniform velocity. Thus with high magnetic strength, there are essentially three laminar regions, one on each side near the walls and a central core.

Parallel Discussion

In the hydromagnetic flow with normal field, the magnetic vector generates an electromagnetic stress just opposite the direction of flow in addition to the viscosity stress of the fluid. Against the two is the pressure gradient to motivate the flow. When flow becomes turbulent, there arises the momentum-exchange stress to equalize the velocity in a conduit. There is exactly the same number of counterparts in the plastic flow; but in place of electromagnetic stress, there is the yield stress created by plasticity.

In order to make sensible comparison of the elementary stresses, we have to look closer to each one of them. In hydromagnetic flow, the retarding force on unit mass is $\sigma B^2\mathbf{u}$, but the net effect on shearing stress is λ, the ratio of magnetic pressure to dynamic pressure. λ is proportional to B^2/u^2 and that shows it is very sensitive to the changes of magnetic strength and the velocity of flow. In plastic flow, the yield value is produced by the proportion of dispersion phase to set up a fixed orientation in structure. It varies with the proportion of the suspended material and only when a certain limit is exceeded "to force the particles into contact with each other, a measurable force is required to cause them to slide over one another, thus exhibiting a yield value." Theoretically, flow is possible, when the local shearing stress is above the yield value. However, seepage with flow of fluid squeezed through the dispersed phase will occur whenever the local shearing stress is very intense but still below the yield value. In a word, both the electromagnetic stress and the yield value tend to lock the fluid particles together, to increase the net shearing resistance, to suppress the turbulence, and thereby to increase the thickness of the laminar boundary layer near solid walls and to extend the range of laminar regime.

The most interesting point is the interaction of the two sets of stresses in the flows. In laminar stage, the turbulent shear of momentum exchange is absent, and the flow pattern is completely controlled by viscosity of the fluid, the pressure drop and the extraneous factor, i.e., interaction of magnetic field and fluid flow represented by λ in hydromagnetic flow, and the effective yield value τ_y modified by the squeezing effect of the actual stress in the plastic flow. Thus we can see, in small velocity, the three stresses may be well balanced to produce very regular relations. This is proved by the fact that Babbitt and Caldwell in plastic flow and Hartmann in hydromagnetic flow provided reliable analyses in the laminar regime. In the turbulent flow, on the other hand, the viscous shear may become negligible, and in its place we have the turbulent stress which has to be balanced with pressure gradient and the electromagnetic or yield stress. Both of the latter have the tendency to bring the flow back to laminar stage. Since both vary with shear intensity the result is the creation of three laminar regions and two transition zones even in high-velocity flow, as described previously [2], [5], [13].

For the transition, Babbitt takes the critical Reynolds number of 2000 and 3000 [1, 2] and incorporates the yield value in calculating the critical velocities in the plastic flow. For hydromagnetic flow, Harris [7] found the friction factor f of 0.035 to 0.044. Anyway, the transitions in these two flows are much more complicated than the pure fluid flow. First, all the four basic stresses are acting at the same time with different proportions in different locality and time. Secondly, the parameter for critical change is not a single, but may contain several factors, which have not been determined. Thirdly, both λ and τ_y have long range of variation. For instance, τ_y may be negligible in diluted flow, but very high value in concentrated mixture. Similarly, the turbulence can be completely damped out by high magnetic field.

Conclusion

The hydromagnetic and the plastic flows both are extending the fields of applications in industries and scientific enquiries and more active research is on the way. Both are closely related to hydrodynamic flow but with different inherent characteristics. In laminar stage they have regular relations with all factors concerned. However, in turbulent stage, both have a central core with uniform velocity flanked by transition zones and laminar layers. Both have four basic stresses in action and interaction. Finally in both flows, the transition is very complicated in nature and nothing in the shape of hydrodynamic transition can be found out.

The parallel analysis presented here may reveal the common demand in future research such as (a) determining the shearing stress, (b) the velocity distribution in a cross section, (c) the turbulence variations, (d) heat exchange measurements, and (e) friction factors. In the application, both may be used in (a) controlling the transition, (b) changing the flow pattern, (c) reducing velocity fluctuations and (d) determining the most economical flows.

REFERENCES

[1] Babbitt, H. E., and Caldwell, D. H., Laminar flow of sludges in pipes with special reference to sewage sludge, *Univ. Illinois Bull.* 37, No. 12 (1939).

[2] Babbitt, H. E., and Caldwell, D. H., Turbulent flow of sludge in pipes, *Univ. Illinois Bull.* 38, No. 13 (1940).

[3] Bershader, D., "The Magnetohydrodynamics of Conducting Fluids." Stanford Univ. Press, Stanford, California, 1959.

[4] Brown, G. C., "Basic Data of Plasma Physics." Wiley, New York, 1959.

[5] Chou, T. L., Resistance of sewage sludge to flow in pipes, Paper 1780, *J. Sanit. Eng., Div. Am. Soc. Civil Engrs.* (1958).

[6] Elliott, D. G., A two-fluid magnetohydrodynamic cycle of nuclear-electrical power conversion, Jet Propulsion Lab. Tech. Rept. No. 32-116 (June 30, 1961).

[7] Harris, L. P., "Hydromagnetic Channel Flows." Wiley, New York, 1960.

[8] Ismail, H. M., Turbulent transfer mechanism and suspended sediments in closed channels, *Trans. Am. Soc. Civil Engrs.* 117, 409 (1952).

[9] Kursunoglu, B., "Principle of Plasma Dynamics." Univ. of Miami Press, Miami, Florida, 1961.

[10] Landshoff, R. K. M., "Magnetohydrodynamics." Stanford Univ. Press, Stanford, California, 1957.

[11] Metzner, A. B., and Reed, J. C., Flow of non-Newtonian fluids, correlation of laminar, transition and turbulent flow regions, *J. Am. Inst. Chem. Eng.* 1, 434 (1950).

[12] Metzner, A. B., Flow of non-Newtonian fluids, *in* "Handbook of Fluid Dynamics" (V. L. Streeter, ed.). McGraw-Hill, New York, 1961.

[13] Murgatroyd, W., Experiments of magneto-hydrodynamic channel flow, *Phil. Mag.* 44, 1348-1354 (1953).

[14] Vanoni, V. A., and Nomicos, G. N., Resistance properties of sediment-laden streams, *Trans. Am. Soc. Civil Engrs.* 125, 1140 (1960), and discussion by Chou, T. L., p. 1168.

[15] Wilkinson, W. L., "Non-Newtonian Fluids." Pergamon Press, New York, 1960.

[16] Wood, P. I., Forrest, D. L., and Wilner, B. M., SNAP-8, The first electric propulsion power system, Am. Rocket Soc. Space Flight Report to the Nation, New York Coliseum, October 9-15, 1961.

SUBJECT INDEX

3-

1